Bringing the Sun Down to Earth

Designing Inexpensive Instruments
for Monitoring the Atmosphere

David R. Brooks

Bringing the Sun Down to Earth

Designing Inexpensive Instruments
for Monitoring the Atmosphere

Springer

Author
Prof. David R. Brooks
Institute for Earth Science Research and Education
PA, USA

ISBN: 978-1-4020-8693-9   e-ISBN: 978-1-4020-8694-6

Library of Congress Control Number: 2008931286

Cover photo: Sunset over Manama, Bahrain, December 2004. The orange-red sky over the
capital city of this small island nation in the Arabian Gulf (the name preferred in the Middle East
for this body of water) is due to scattering of sunlight caused by urban pollution and wind-blown
sand and marine spray.

Printed on acid-free paper

9 8 7 6 5 4 3 2 1

springer.com

Preface

In 1998, my colleague, Forrest Mims, and I began a project to develop inexpensive handheld atmosphere monitoring instruments for the GLOBE Program, an international environmental science and education program that began its operations on Earth Day, 1995. GLOBE's goal was to involve students, teachers, and scientists around the world in authentic partnerships in which scientists would develop instrumentation and experimental protocols suitable for student use. In return, data collected by students and their teachers would be used by scientists in their research. This kind of collaboration represented a grand vision for science education which had never before been attempted on such a scale, and we embraced this vision with great enthusiasm.

Between 1998 and 2006, Forrest Mims and I collaborated on the development of several instruments based on Mims' original concept of using light emitting diodes as spectrally selective detectors of sunlight, which was first published in the peer-reviewed literature in 1992. These instruments have evolved into a set of tools and procedures for monitoring the transmission of sunlight through the atmosphere, and they can be used to learn a great deal about the composition of the atmosphere and the dynamics of the Earth/atmosphere/sun system. If measurements with these instruments are made properly, they have significant scientific value, as well.

For most GLOBE Program protocols, the original vision of scientists, teachers, and students collaborating in published research was never realized, but atmospheric science was an exception that led to several peer-reviewed publications about the instruments we developed and their science applications. These kinds of activities require patience because they can take years to bear fruit, even for "professionals." For example, when a completed paper that may represent several years of work is submitted to a peer-reviewed journal, it may not appear in print for yet another year or more.

Sadly, the GLOBE Program and its sponsors lost patience with this process, and GLOBE has now abandoned its original vision of engaging students and teachers around the world in authentic science. Nothing would please me more than for the readers of this book to resurrect this vision in their own schools, homes, and communities.

My experience in working with students and teachers around the world has demonstrated that it is very rare to find individuals with the skills required to design and build their own instruments. In my view, this is a problem within science education that seriously limits students' abilities to understand the world around them. Today's students have grown up in a world dominated by digital technology. However, the world is a fundamentally analog place and must be understood on those terms.

Atmospheric and Earth sciences rely heavily on observation, and only by immersing ourselves in observations and measurements can we understand the atmosphere and the dynamic processes that drive weather and climate.

These kinds of activities are not just for an academic elite. Our nation is struggling to retain and educate a competitive world-class science and engineering workforce, but this workforce must also include a technical support infrastructure that requires practical as well as academic skills. A great deal of attention is being paid to the academic side of this equation, but at least in my perception, very little to the practical side.

Although this book is based in large part on what I have learned by working with K-12 students and their teachers around the world, I know that the audience extends into colleges and universities. I am often contacted by professors who have discovered that their students, even those who are candidates for advanced degrees in science and engineering, have no idea about how to design, build, calibrate, maintain, and use the instruments that are essential to their careers. (A few years ago I had a student assistant who was a sophomore mechanical engineering major at Drexel University and who had never seen an analog panel meter.) For that audience, this book should provide an valuable introduction to essential skills. After all, the process of designing instruments and experiments remains fundamentally the same regardless of whether an instrument costs $20 or $20,000. It certainly makes sense to develop and practice new skills with $20 instruments rather than with their more expensive counterparts!

This book is not intended for use just in formal education settings. It provides a great deal of material for science projects that can be conducted by anyone interested in their environment. Traditionally, "environmental science" projects have not fared as well in science fairs as projects in other areas. Perhaps this is due to a lack of information about making authentic quantitative measurements. If so, this book, in which I have tried to strike a balance between science and practical design matters, should help!

There are several places in this book in which I have discussed prices of instruments and components, in order to give readers a better idea of my definition of "expensive" and "inexpensive." Although the relative price comparisons should hold in the future, the absolute prices are in 2008 US dollars and, of course, may not be directly applicable in the future.

Although it is fair to describe these instruments as relatively inexpensive compared to their commercial counterparts, it may not be inexpensive or easy to build just one or two. There are almost always cost advantages to buying components in quantity.

It is inevitable that there are some practical construction details that I should have included but did not—the book was conceived more as an "idea manual" than a "construction manual." Building these instruments may require specialized hardware and tools. In my own shop I have all the equipment I need for the required mechanical and electronic work. I have designed and had manufactured custom enclosures and printed circuit boards for some of these instruments, but these undertakings are practical only when ordering items in quantity.

In some cases, kits of parts and detailed construction instructions may be available through *Institute for Earth Science Research and Education.* However, I cannot predict what kinds of assistance will be available when you read this book. I invite you to search for me and *IESRE* on the Web and contact me to find out about current activities.

There are many facts about the sun and the Earth/atmosphere system presented in this book, especially in Chapter 2, and I have not attempted to provide specific references for all of them. If there are errors, I take full responsibility for them. There are, of course, many online sources that can be used to check facts and obtain more information about specific topics.

Unless noted otherwise, all figures and tables represent my own work. All figures in the text are printed in black and white. However, in a few cases where I believed that color versions were necessary to render the figures understandable, figures are reproduced in color plates at the end of the book, with the same chapter-specific figure numbers.

Finally, note that although the URLs for online sources given occasionally in this book were valid at the time the manuscript was written, there is no guarantee that they will be available in the future.

Acknowledgements

Some of the material in this book is based on work made possible by grants and other support from the National Science Foundation (for work with the GLOBE Program), the National Aeronautics and Space Administration's Goddard Space Flight Center and Langley Research Center, and the National Oceanic and Atmospheric Administration.

Anyone who is interested in the kinds of measurements and instruments described in this book is deeply indebted to the groundbreaking and continuing work by Forrest M. Mims III, the inventor of LED-based instruments for atmospheric measurements, and Co-Principal Investigator with me for eight years on the GLOBE Program's Aerosols and Water Vapor Monitoring Projects. Mr. Mims has been a professional inspiration, a valued colleague, and a personal friend since the beginning of our work together in 1998.

I also want to thank Wade Geery, a teacher at Arrie Goforth Elementary School in Norfork, Arkansas, USA, and Richard Roettger, a teacher at Ramey School, Ramey, Puerto Rico, for their ongoing commitment to doing real science with their students and for collecting and reporting atmospheric data over several years, some of which appear in this book.

Finally, I need once again to thank my wife, Susan, for proofreading this manuscript prior to its submission. Her patience is especially noteworthy considering that this was the third book manuscript I asked her to read within a period of less than a year and a half.

David R. Brooks

Institute for Earth Science Research and Education
Worcester, Pennsylvania, USA
February, 2008

Contents

1. Introduction

For many years, the principles of inquiry-based science activities have spread throughout the science education community, and the "hands-on" mantra has gained the status of a cliché for education reform. Nevertheless, the transition from cliché to reality remains elusive, at least partly because of the gap between the worlds of working scientists, classroom educators, and students. In an educational environment increasingly driven by the use of standardized testing to assess the performance of schools, teachers, and students, and with rewards and penalties based on scores on these tests, it is difficult for even the most diligent educators to involve their students in doing real science in partnerships with scientists. Correspondingly, it is also difficult for scientists to provide full access to and participation in a research environment that can seem unapproachable to a nonscientist.

The purpose of this book is to bridge these gaps in one particular area—the science associated with interactions between solar radiation and the Earth/atmosphere system. Although this topic may seem esoteric, in fact it is very approachable by teachers, students, and others spanning a wide range of ages and interests, and is a necessary component of any curriculum that meets national science education standards [National Research Council, 1996].

This topic is important to all of us because sunlight is the ultimate fuel source driving the Earth/atmosphere "engine." Most obviously, solar energy and its daily and seasonal variability provide the driving force for weather and climate. Even the youngest students are taught to make simple weather observations and to answer questions such as, "Has it been raining?" "How cloudy is it this morning?" "How hot will it be today?" Basic qualitative observations can lead to quantitative measurements and more sophisticated questions: "How much rain have we had this month, and is that more or less than average?"; "What is the percentage of cloud cover and what kinds of clouds are present?"; "Is it getting warmer where we live?"; "Why does the atmosphere appear hazy and what does this have to do with weather and climate?"

These kinds of questions are interesting to students and others because they are so closely bound to our daily experiences and relate to issues of climate change that will dominate Earth science in the 21st century. Activities that help answer these questions serve as essential

starting points for building a scientifically literate society that understands, for example, the potential for and consequences of human-induced
climate change, and which values the investments required to lead students
to pursue professional careers in the "STEM" disciplines—science, technology, engineering, and mathematics. However, these activities must be
chosen with care to make sure that they provide reliable results and that
those results are used appropriately.

As a first step toward studying sun/Earth/atmosphere interactions,
this book addresses how these interactions maintain Earth and its atmosphere in a radiative balance that supports life as we understand it, and
how to measure the effects of those interactions. These processes are
illustrated conceptually in Figure 1.1. Some version of this image is
invariably found in the Earth science texts often used in 8th and 9th grade
courses [*e.g.*, Allison *et al.*, 2006]. There is a great deal of science
embedded in such images, most of which no doubt remains a mystery to
non-specialists, including students *and* their teachers.

Earth and its surprisingly thin and fragile blanket of atmosphere
(with a thickness equal to about 1/100 of Earth's radius) form a dynamic,
interconnected system. Incoming sunlight is reflected, scattered, and
absorbed by Earth's atmosphere and its surface. Different surfaces,
including cloud top "surfaces," reflect and absorb sunlight in different
ways. Oceans absorb almost all the radiation that falls on them. New snow
reflects nearly all radiation. Whatever the surface, absorbed radiation is
ultimately re-emitted as longwave thermal radiation, as indicated by the
red arrow in Figure 1.1.

Earth's yearly journey around the sun and its daily rotation about
an axis that is tilted relative to its orbital plane keeps Earth and its atmosphere in a constant state of flux that is driven by diurnal and seasonal
cooling and heating. On average, this ongoing process of absorption and
re-emission keeps the Earth/atmosphere system in the radiative balance
that is required by basic physical laws.

The details of the processes summarized in Figure 1.1, are what
fascinate atmospheric and Earth scientists. Although the explanations can
sometimes be complicated, atmospheric and Earth sciences still rely heavily
on observation. Many observations and quantitative measurements of the
atmosphere are not difficult to understand and are well within the capabilities of students, teachers, and anyone else who is curious about their
environment.

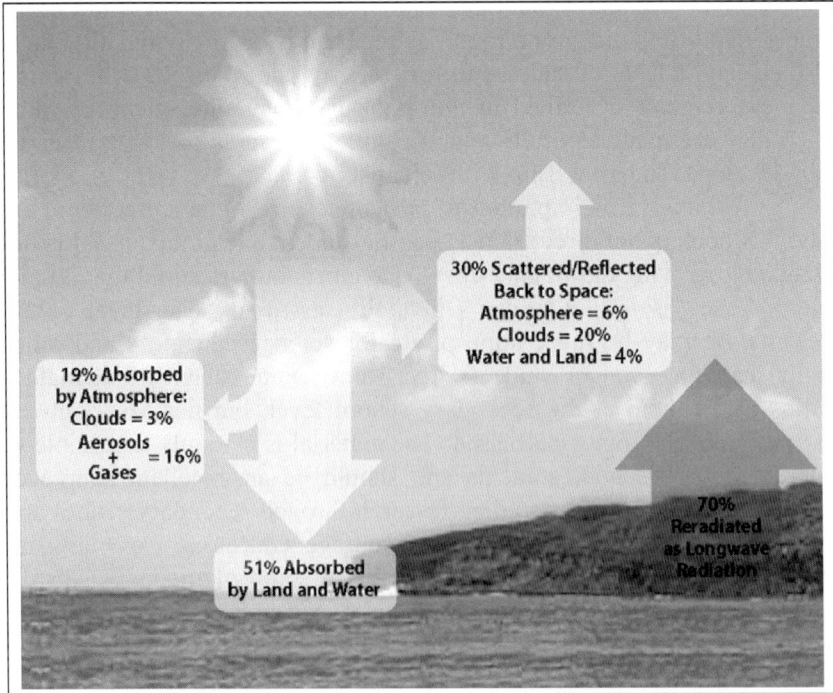

Figure 1.1 (see color plates). Schematic representation of Earth's radiative balance [Graphic by Vivek Dwivedi, NASA Goddard Space Flight Center].

Even the simplest measurements require an understanding of the principles of scientific research and of how measuring instruments work. The principles and challenges of designing and building scientific instruments, including dealing with their calibration, interpretation, stability, and reliability, are the same regardless of whether those instruments cost $20 or $20,000 and should be applied to all activities in appropriate ways.

Although it may be tempting to conclude that the accuracy of observations and measurements made "just" for educational purposes is not particularly important, this is a mistake! Without observations and measurements that make sense relative to accepted scientific standards, it is easy for students to become discouraged. Without reasonable data quality standards, scientists will not be motivated to participate in the learning process. But, when scientists, teachers, and students collaborate to design experiments, including building their own instruments as described in this book, participants are engaged in an authentic science experience that confers true ownership of the data and benefits all stakeholders.

In this context, examining sunlight and its interaction with Earth's atmosphere offers many opportunities for "hands on" activities that meet all the requirements of state-of-the-art science education. In addition, all the measurements described in this book have genuine scientific value when they are made carefully and, ideally, are used as part of a research plan developed in partnership with scientists.

Although it is important to present science in an age-appropriate way, this book is not directed at a specific student age group, and it is not intended to provide thoroughly developed curriculum material that is suitable for use "as is" by teachers in their classrooms. The level of the presentation may be more appropriate for science educators and other adults than for younger students. It assumes some science and mathematics background at the secondary school level, but not a specialized knowledge of the topics discussed. The material is certainly accessible to older secondary school students and should be an excellent source of science projects and other student research through secondary school and beyond. Hopefully, this book will encourage educators, especially, to develop their own science interests, to build some of their own instruments for monitoring the atmosphere, and to transmit their enthusiasm to their students. The book could easily serve as a source for undergraduate and even graduate courses in environmental or atmospheric science or environmental engineering.

The approach taken in this book has been developed specifically because it is so easy simply to buy equipment for collecting science data and use it without giving much thought to the underlying design principles. This may seem reasonable and expedient in some circumstances, but it removes an important layer of understanding from the science process. Because of our society's increasing dependence on digital technology, this book (in spite of its embracing technology as appropriate) will hopefully serve as a reminder that the physical world is an analog place. Anyone who understands this fact and has the skills to deal with it will continue to be in great demand in science and many other careers!

The book is divided into three basic sections. It begins with an introduction to the sun, Earth's atmosphere, and some of its constituents. The second section presents a discussion of the principles behind full-sky and direct sun measurements and the science that is reasonably accessible with relatively inexpensive instruments.

The third section, starting with Chapter 4, describes how to design and use atmospheric monitoring instruments that are both inexpensive and reliable. Supplementary materials in appendices delve more deeply into some of the mathematical and technical issues, but they are not essential for understanding and applying the material in the book.

2. Earth's Sun and Atmosphere

Chapter 2 provides some facts about the sun and Earth's atmosphere. Equations are presented that show how the Earth/atmosphere system is maintained in equilibrium with the radiation arriving from the sun, and how greenhouse gases in the atmosphere affect that balance.

2.1 Earth's Sun

Stars generate huge amounts of energy through the process of nuclear fusion, in which hydrogen atoms are converted into helium atoms. Earth's sun, an unremarkable medium-sized star, produces a total power P of about 3.9×10^{26} watts (W).[1] This power is radiated into space uniformly in all directions. At planetary ditances, the sun looks like a point source of radiation and fundamental physical laws tell us that the intensity of a point source of radiation decreases as the inverse square of the distance from the source. The solar constant S_o is defined as the power per unit area of solar radiation falling on the surface of an imaginary sphere of radius R around the sun:

$$S_o = P/(4\pi\, R^2) = P/(4\pi\cdot150,000,000,000^2) \approx 1370 \text{ W/m}^2 \qquad (2.1)$$

where R is the average Earth/sun distance (1 astronomical unit or 1 AU), about 1.5×10^{11} m. The solar output actually fluctuates a little as a result of disturbances on the sun's surface. More importantly, the solar "constant" varies by about ±3.4% during a year because Earth is in a slightly non-circular orbit around the sun. The maximum and minimum values occur at perihelion and aphelion—the minimum and maximum Earth/sun distances:

$$S_{max} = S_o/(1 - e)^2 = S_o/(0.983)^2 = 1417 \text{ W/m}^2 \text{ (at perihelion)} \qquad (2.2a)$$
$$S_{min} = S_o/(1 + e)^2 = S_o/(1.017)^2 = 1324 \text{ W/m}^2 \text{ (at aphelion)} \qquad (2.2b)$$

---

[1] A list of symbols, organized by chapter, is given in Appendix 1.

where e is the eccentricity—a dimensionless measure of the departure of an orbit from a circle with a value between 0 and approaching 1. Earth's eccentricity varies slowly, with a period of about 100,000 years. The current value is about 0.0167. Earth is closest to the sun, at perihelion, in early January, so this is when maximum solar radiation reaches Earth.[2] The minimum amount of radiation is received at aphelion, 6 months later.

Light and heat are the obvious perceptible components of solar radiation, but the energy distribution of solar radiation is much more complex than that. The sun's energy, like that generated by other stars, is distributed over a broad range of the electromagnetic spectrum, following well-known physical laws. It behaves approximately like a "blackbody," a perfect radiator and absorber, at a temperature of about 5,800 K— "approximately" because of electromagnetic activity and other processes constantly taking place within the sun's interior and on its surface. Its maximum output is in the green-yellow part of the visible spectrum, around 500 nm. Not surprisingly, this is near the maximum sensitivity of the human eye.[3]

Astronomers classify stars using a series of letters based on their equivalent blackbody temperatures, as shown in Table 2.1. Earth's sun has an absolute temperature of about 5,800 kelvins (K), which places it near the upper end of type G stars, near the middle of the range of star temperatures.

The radiation leaving the sun on its journey through the solar system is not a completely smooth function of wavelength. Figure 2.1 shows the extraterrestrial solar radiation—what an observer would see from a vantage point just above Earth's atmosphere—obtained from the SMARTS2 model. For reference, the distribution of blackbody radiation at 5,800 K as a function of wavelength is also shown. Appendix 2 provides more details about the equation for Planck's law, one of the most famous in the history of physics, needed to generate the blackbody radiation curve.

---

[2] Although many people living in the northern hemisphere believe the sun must be closer to Earth during the northern hemisphere summer, this is not true!

[3] It is reasonable to conclude that the human eye has evolved to respond optimally to solar radiation. The fact that other animals respond differently, for example, by being able to "see" ultraviolet or thermal radiation, suggests many fascinating questions that are beyond the scope of this book.

Table 2.1. Star type classifications by temperature
[Cannon and Pickering, 1912].

Spectral letter	Temperature range (K)	Stellar color
O	>30,000	Blue
B	10,000–30,000	Blue
A	7,500–10,000	Blue-white
F	6,000–7,500	White
G	5,000–6,000	Yellow-white
K	3,500–5,000	Orange
M	<3,500	Red

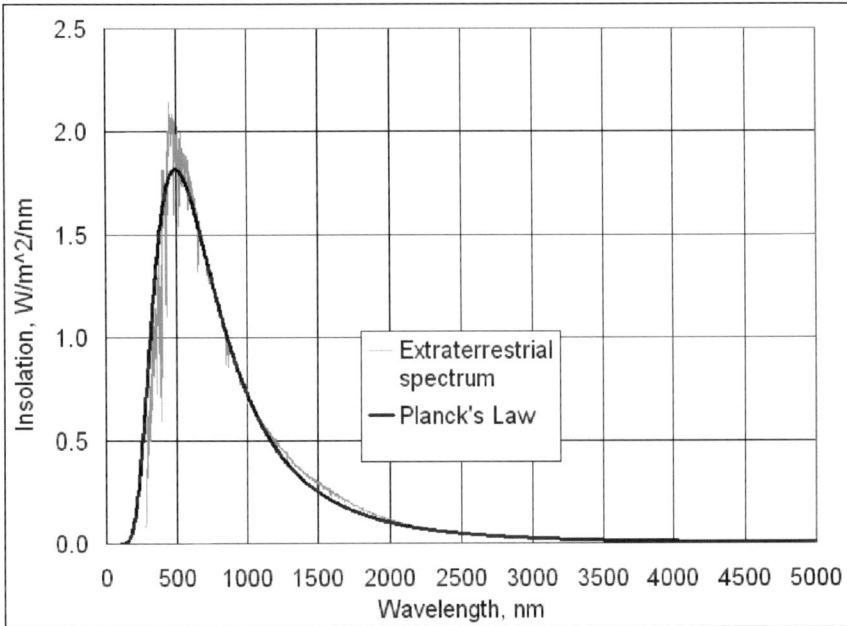

Figure 2.1. The extraterrestrial spectrum [from SMARTS2 model, Gueymard, 1995] with Planck's law blackbody radiation for a temperature of 5,800 K superimposed.

2.2 Earth's Atmosphere

Earth's size, density, and distance from the sun in a nearly circular orbit have produced a fortuitous set of circumstances for supporting life as we understand it. Gravity keeps in place an oxygen-rich atmosphere. The average surface temperature (about 16°C), as controlled by the solar constant and the atmosphere, allows water to exist naturally in all its three phases—solid, liquid, and gas. Although it is almost certainly a mistake to assume that Earth provides a unique environment for the development of life in the universe (considering that scientists continue to find primitive life in Earth environments that seem too hostile to support life), it is certainly true that conditions supportive of a permanent oxygen-rich atmosphere and abundant water for most of Earth's long history have made possible the development of advanced forms of life as we understand them.

The atmosphere consists of gases, predominantly nitrogen and oxygen. Table 2.2 gives the composition of pure dry air near Earth's surface.

Table 2.2. Composition of pure dry air near Earth's surface
[http://www.arm.gov/docs/education/backgroundcompositionatmos.htm].

Gas	Percent by volume (dry air)	Cumulative percent by volume
N_2	78.08	78.08
O_2	20.95	99.03
Ar	0.934	99.964
Other trace gases	0.036	100.000

The actual atmosphere contains other naturally occurring and anthropogenic (human-produced) components, as shown in Table 2.3.

Table 2.3. Trace gases in the atmosphere
[American Chemical Society, 2000].

Component	Approximate percent by volume and parts per million (ppm)
Water vapor (H_2O)	0–4%
Carbon dioxide (CO_2)	0.037% (370 ppm)
Methane (CH_4)	0.00017% (1.7 ppm)
Nitrous oxide (N_2O)	0.00003% (0.3 ppm)
Ozone (O_3)	0.000004% (0.04 ppm)
Aerosols (liquid and solid particles)	0.000001–0.000015% (0.01–0.15 ppm)
Chlorofluorocarbons (CFCs)	0.00000002% (0.0002 ppm)

Although the amounts of these trace components may seem *very* small, except perhaps for water vapor, their effects on the Earth/atmosphere system are profound. Their contribution is popularly referred to as the "greenhouse effect." The basic mechanism is that the atmosphere is transparent to most incoming solar radiation over a wide range of wavelengths. This energy heats Earth's surface, which then re-radiates radiation in the infrared, some of which is absorbed by molecules in the atmosphere. This raises the temperature of those molecules, which warms the Earth/ atmosphere system.

However, a greenhouse turns out to be far from a perfect analogy for what happens in the atmosphere. Experiments conducted a century ago have shown that, even when a laboratory greenhouse is made from materials that are highly transparent to emitted thermal radiation, such as rock salt, it still gets warm inside [Wood, 1909]. This is due to the fact that the air and surfaces within a closed greenhouse are warmed by the re-emitted thermal radiation. Air temperatures rise because there is no transport of air through the greenhouse. However, the *effect* on the Earth's atmosphere or on the air trapped within a greenhouse is the same—the air gets warmer. On a global scale, the greenhouse analogy is not too bad because air, whether it is transported or not, is still trapped within the Earth/atmosphere system. In any event, the "greenhouse" image is so embedded in the popular understanding of global warming that there is little point in trying to do away with this term.

It is not difficult to quantify the overall greenhouse effect on the Earth/atmosphere system. Consider the solar radiation striking Earth. The area of Earth's disk as viewed from space is

$$\text{area} = (\pi r^2)\ \text{km}^2 \tag{2.3}$$

where r is Earth's radius, about 6,378 km at the equator.[4] Energy from the sun, equal (on average) in intensity to the solar constant S_o, is intercepted by Earth's disk, so the total energy incident on Earth is

$$\text{incident energy} = (\pi r^2)S_o \tag{2.4}$$

Not all of the incident solar energy is absorbed; a portion is reflected back to space, as determined by the average reflectivity, or albedo A, of the Earth/atmosphere system:

$$\text{absorbed energy} = (\pi r^2)S_o(1 - A) \tag{2.5}$$

As viewed from space, the albedo of the Earth/atmosphere system is about 30%. Therefore, about 70% of the incident solar energy is absorbed by the Earth/atmosphere system.

---

[4] The Earth is not a perfect sphere, but bulges slightly at the equator. Its polar radius is about 6,357 km.

Basic physical laws require that bodies must, on average, be in radiative equilibrium. That is, whatever energy is absorbed must be re-emitted in some form. The solar energy striking Earth's disk as viewed from space is re-emitted as thermal radiation by the surface of the entire planet, a total area of $4\pi r^2$, in proportion to the fourth power of the absolute temperature,[5] as described by the Stefan-Boltzmann law:

$$\text{emitted energy} = (4\pi r^2)\sigma T^4 \qquad (2.6)$$

where the Stefan-Boltzmann constant $\sigma = 5.67 \times 10^{-8}$ W/(m^{2}K^4). On average, the absorbed energy must equal the emitted energy:

$$(\pi r^2)S_o(1-A) = (4\pi r^2)\sigma T^4 \qquad (2.7a)$$

or, simplified,

$$S_o(1-A) = 4\sigma T^4 \qquad (2.7b)$$

Solving for T yields:

$$T = [S_o(1-A)/(4\sigma)]^{(1/4)} = [1370 \cdot (1-0.3)/(4 \cdot 5.67 \times 10^{-8})]^{(1/4)} = 255 \text{ K} \qquad (2.8)$$

0°C is about 273 K, so the temperature at which the Earth/atmosphere system is in radiative equilibrium as viewed from space is about -18°C! This is very cold from a human perspective—far below the freezing point of water. But the average Earth surface temperature is actually about 16°C, a much more pleasant value from a human perspective. The greenhouse effect accounts for the difference of about 34°C, by absorbing emitted thermal radiation and re-radiating some of it back to Earth's surface.

A very simple way of illustrating the greenhouse effect is to modify the radiative balance equation:

$$S_o(1-A) = 4\sigma T^4(1-x) \qquad (2.9)$$

where "x" is a "greenhouse factor," with a value between 0 and 1, which provides a measure of the net effect of radiation from Earth's surface which is absorbed by the atmosphere and re-radiated down to Earth's

---

[5] A "Kelvin" (not a "degree Kelvin") is the metric unit for absolute temperature. "Absolute zero" is 0 K, or -273.15°C.

surface, rather than out to space. For x = 0, there is no absorption and no greenhouse effect. For Earth, a value of x = 0.4 produces an equilibrium temperature for the Earth/atmosphere system of about 16°C. Values closer to 1 lead to a "runaway greenhouse effect," such as exists on Venus, resulting in very high surface temperatures.

Equation (2.9) is the basis of a very simple climate model that you can use to design your own planet. The model variables include different distributions of surfaces, including cloud "surfaces," each with its own albedo, the planet/sun distance, and the greenhouse factor. The model output is the average surface temperature of the planet. Although it is important not to expect too much from such a simple model, it nonetheless illustrates some important general principles that explain why Earth is the way it is. This model is discussed in Appendix 3.

As noted above, the greenhouse effect is essential to support advanced life on Earth. Although there have been large variations in Earth's climate during its long history, the rates of global change have generally been slow enough to allow many life forms to adapt.[6] However, even seemingly small changes in the concentrations of the more potent greenhouse gases can disrupt entire ecosystems on time scales that are short relative to historical precedents. If changes occur too rapidly, some species may not have enough time to adapt. This is a primary source of concern about human-induced changes in Earth's delicately balanced greenhouse.

Table 2.4 shows the relative effectiveness of some greenhouse gases at trapping infrared radiation in Earth's atmosphere. CO_2 is arbitrarily given an effectiveness of 1. Water vapor is the most prominent greenhouse gas, and its relatively low effectiveness is offset by its high concentration in the atmosphere. Carbon dioxide is an important greenhouse gas even at the concentration shown in Table 2.3. CO_2 has both natural and anthropogenic sources, and is of great concern because steadily increasing CO_2 levels due to the burning of fossil fuels since the start of the Industrial Revolution are generally believed to be warming Earth's climate more quickly than at any time in the recorded past.[7]

---

[6] Well-documented "mass extinctions" of life many millions of years ago are generally thought to be due to unique events, such as a comet or asteroid colliding with Earth, which caused very abrupt climate changes.

[7] "Recorded past" includes conditions inferred from indirect evidence of past climates found in ice cores, for example.

Table 2.4. Relative effectiveness of greenhouse gases
[American Chemical Society, 2000].

Greenhouse gas	Relative effectiveness
Carbon dioxide (CO_2)	1 (arbitrarily assigned)
Methane (CH_4)	30
Nitrous oxide (N_2O)	160
Water (H_2O)	0.1
Ozone (O_3)	2,000
Trichlorofluoromethane (CCl_3F)	21,000
Dichlorodifluoromethane (CCl_2F_2)	25,000

The most notorious greenhouse gases belong to the family of chemicals known as chlorofluorocarbons (CFCs). These manmade chemicals have no natural sources and have entered the atmosphere solely as the result of human industrial activity. Because of their chemical stability, CFCs were used as coolants and as propellants in spray bottles. In addition to being potent absorbers of thermal radiation, and even at what appear to be miniscule concentrations, CFCs are primarily responsible for dramatic seasonal reductions in concentrations of stratospheric ozone (the now-famous "ozone holes" [NASA, 2001]).[8] The high relative effectiveness of CFCs is due to the fact that chlorine atoms, when separated from their compounds by high does of UV radiation in the stratosphere, act as catalysts in converting ozone molecules to oxygen molecules. A single chlorine atom can participate in tens of thousands of such reactions.

This is a problem because stratospheric ozone is the "good ozone" that protects Earth's surface by absorbing ultraviolet (UV) radiation that can have harmful effects on humans and other organisms. (The "bad ozone" near Earth's surface, which is a pollutant that can cause serious health and environmental problems, is not affected in this way because it is produced locally as a result of ongoing photochemical processes at Earth's surface.)

---

[8] There were not really "holes" in the stratospheric ozone, but observed seasonal reductions in ozone levels were far greater what had been predicted or seen previously.

3. Measuring Atmosphere and Surface Properties

Chapter 3 provides an overview of instrumentation for measuring sunlight, including full-sky and direct sunlight measurements of broadband and spectrally selective radiation. It examines what kinds of science can be done with relatively inexpensive instruments and relates measurements accessible to non-specialists to a global view of atmospheric science.

3.1 The Distribution of Solar Radiation

Monitoring Earth's atmosphere is a challenging task. In industrialized countries, there are well-established networks of instruments to monitor the atmosphere close to Earth's surface. Some stations are used for scientific purposes, but most serve the primarily regulatory function of monitoring a specific set of air quality indicators as mandated by government agencies. (In the U.S., the "criteria pollutants" on which the Air Quality Index is based, as established by the Environmental Protection Agency, are SO_2, NO_2, O_3, CO, and particulates.) Although these quantities are certainly of scientific interest, they are only local measurements and represent only a small subset of important atmospheric measurements.

In order to understand how the Earth/atmosphere system works, it is necessary to understand the entire atmosphere, not just the part close to Earth's surface. Often, the vertical distribution of gases and particles in the atmosphere is not known precisely, so the total amounts of those gases and particles in a column of atmosphere cannot be determined from measurements made just at Earth's surface—essentially at the bottom of the atmosphere column. Balloons, airplanes, and rockets are all used to perform direct measurements in the atmosphere at altitudes up to and beyond the stratosphere to help understand the vertical distribution of atmospheric constituents. Satellite-based instruments provide global views, but it is difficult to infer surface and column vertical distributions from space-based measurements; such measurements must still be supplemented by ground-based measurements.

An important means of exploring the atmosphere from the ground is to measure the effects of the atmosphere on sunlight transmitted through the atmosphere to Earth's surface. These techniques provide information about the entire atmosphere above the observer, but not about vertical distributions. Still, such upward-looking measurements are especially valuable when they are compared with downward-looking measurements from space. Many of these ground-based measurements are not difficult to make, and are well within the capabilities of students and other non-specialists.

As noted previously, not all solar radiation incident at the top of the atmosphere reaches Earth's surface. An average of about 30% is reflected or scattered back to space by the atmosphere, surface, and clouds. The total amount of solar radiation reaching Earth's surface is called the insolation, or surface solar irradiance. Some radiation is absorbed by the atmosphere and some by Earth's surface. All absorbed radiation is eventually re-emitted, but at much longer wavelengths.

Figure 3.1 shows the spectral distribution of solar radiation reaching Earth's surface for a "standard atmosphere" (a clean and cloud-free atmosphere with specified characteristics) compared with the extra-terrestrial radiation above the atmosphere at the average Earth/sun distance (as shown previously in Figure 2.1). The global, or total, solar radiation at Earth's surface consists of two components: direct radiation from the sun and diffuse radiation from the rest of the sky. The relationship between direct and diffuse radiation under a clear sky depends primarily on the position of the sun in the sky. (This relationship is much more com-plicated under cloudy skies.) The sun in Figure 3.1 is at a zenith angle z of about 48°, giving a relative air mass of about 1.5. Relative air mass m_{air} is a measure of how much atmosphere direct sunlight passes through on its way to Earth's surface. Approximately, $m_{air} \approx 1/\cos(z)$. When the sun is directly overhead, $z = 0°$ and the relative air mass is 1, by definition.

The ratio of diffuse to direct radiation is a strong function of wavelength, especially at ultraviolet wavelengths below about 300 nm. Figure 3.2 shows this ratio starting at 300 nm for the same atmospheric and solar position conditions that were used to generate Figure 3.1.

The data shown in Figure 3.2 offers an important clue about what happens in the atmosphere: molecules scatter sunlight by an amount related to their wavelength. Some of the scattered light is directed back to space and some of it reaches Earth's surface. As a result, the sky itself appears to be a source of light even though all the radiation is actually coming just from the sun. (Beyond the atmosphere, there is no diffuse sunlight. This is why "space" appears black.)

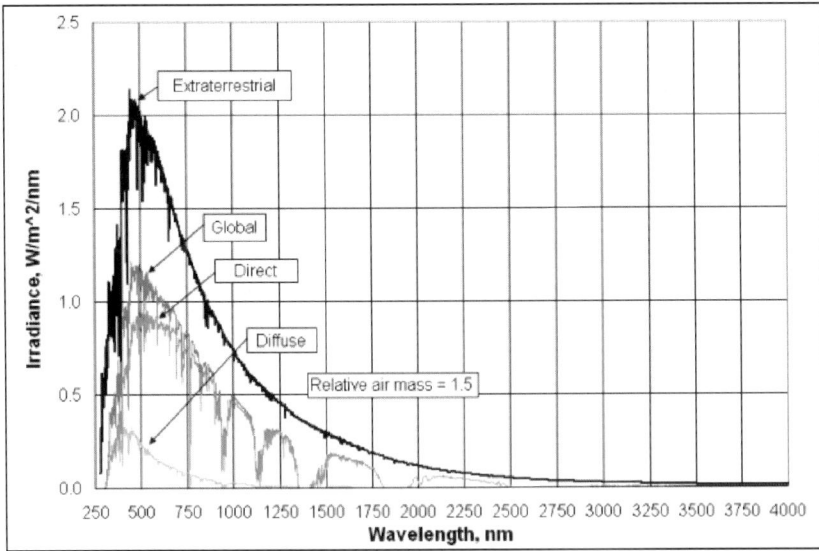

Figure 3.1 (see color plates). Direct, diffuse, and total insolation at Earth's surface for a standard atmosphere and a relative air mass of 1.5.

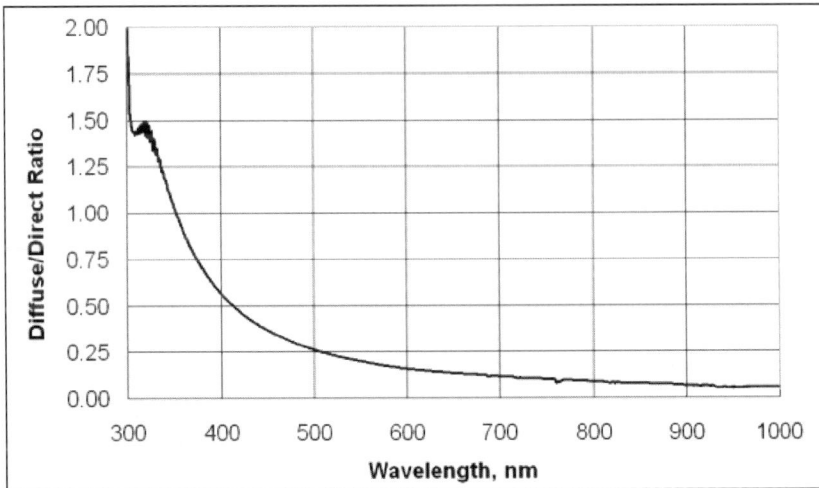

Figure 3.2. Diffuse/direct ratio for solar radiation at Earth's surface, using a standard atmosphere and a relative air mass of 1.5.

Scattering of light by molecules is called Rayleigh scattering, after the British physicist John William Strutt, the third Baron Rayleigh (1842–1919), who first described this phenomenon mathematically. The fact that molecules scatter blue light more efficiently than light having longer wavelengths explains why the sky appears blue. As the sun approaches the horizon through a hazy sky, more and more of the shorter wavelengths of light are scattered out of the line of sight between an observer and the sun, leaving only orange and red light. As particles get larger, scattering becomes less wavelength-dependent. Clouds appear white because the large water droplets in clouds scatter and reflect light uniformly throughout the visible range of wavelengths.

Solar radiation is not just reflected and scattered as it passes through the atmosphere. Some radiation is absorbed by molecules and particles and re-emitted as thermal radiation. The nature of molecular bonds permits the absorption of radiation only at specific wavelengths. As a result, insolation is not just a uniformly diminished version of the extraterrestrial radiation. Instead, there are "holes" in the insolation spectrum caused by molecular absorption.

The largest absorption "holes" in insolation are due to water vapor—hence its importance as a greenhouse gas. Prominent water vapor absorption occurs far out in the infrared part of the solar spectrum, but there are also absorption bands in the near-IR—around 720, 820, and 940 nm, for example, as shown in Figure 3.1.

The second most important absorber in the UV-visible-near-IR regions of the solar spectrum is ozone. There is a bell-shaped ozone absorption band between 210–310 nm (the Hartley band), as well as less prominent bands between 310–350 nm (Huggins band) and 450–850 nm (Chappuis bands). Ozone absorbs almost all UV-C radiation (200–280 nm) and roughly 70% or more of the UV-B (280–320 nm) radiation. There is little atmospheric absorption of UV-A radiation (320–400 nm). The ozone bands are not as deep and wide as the water vapor absorption bands and are not clearly visible at the resolution of Figure 3.1.

In addition to scattering caused by molecules, larger particles suspended in the atmosphere, called aerosols, also scatter sunlight. This is called Mie scattering, after the German physicist Gustav Mie (1868–1957), who first described this phenomenon mathematically. These particles have sizes in the same range as the wavelength of light (~100–1,000 nm), so they scatter light differently than molecules, which are much smaller.

The measurable effects of scattering and absorption enable scientists to develop ground-based measurement strategies based on this important conclusion:

Absorption and scattering by molecules and particles in the atmosphere leave "fingerprints" on the insolation spectrum which provide a means of measuring quantitatively the presence and effects of those molecules and particles.

With this fact in mind, the next step is to determine how best to measure not just total insolation, but also subsets of solar radiation at specific wavelengths.

3.2 Instrumentation Principles for Measuring Sunlight

There are two general categories of instruments used to measure the transmission of sunlight through Earth's atmosphere: instruments that measure radiation from the entire sky, and instruments that measure only direct solar radiation. Within each of these categories, instruments can be further subdivided into those that measure radiation over a broad range of wavelengths and those that measure only specific wavelengths. Each of these four classes of instruments has an important role to play in understanding sun/atmosphere interactions.

3.2.1 Full-Sky Instruments

As their name implies, full-sky instruments need an unobstructed view of the entire sky. Therefore, they need to be installed at sites that have a 360° view of the horizon, without significant obstacles. Corrections can be made for limited horizon obstructions, but the less cluttered the horizon is, the more accurate full-sky measurements will be. Full-sky instruments can be used in places with more restricted visibility, too, but then the interpretation of their data will be much less clear.

An important requirement for full-sky detectors is that they have good "cosine response" to direct sunlight. If sunlight has intensity I_o when the sun is directly above a horizontal surface (a zenith angle of 0°), then the intensity I_z at some other zenith angle z is

$$I_z = I_o \cos(z) \tag{3.1}$$

If an ideal detector on a horizontal surface is illuminated by direct light, then its response should be proportional to the cosine of the zenith angle of the light source. Real detectors do not have perfect cosine response. Cosine response corrections for direct sunlight can be determined and applied under cloud-free skies, but this issue becomes much more complicated under partly cloudy skies, because the relative contributions by direct and diffuse radiation are difficult or impossible to distinguish precisely.

Full-sky instruments are called radiometers or, in the case of broadband solar detectors, pyranometers. Figure 3.3 shows a Model PSP pyranometer manufactured by The Eppley Laboratory, Inc., a widely used "first-class" reference radiometer, as defined by the World Meteorological Organization. This instrument is about 20 cm in diameter. The sensor is under the polished hemispherical glass dome. The glass is specially formulated to transmit solar radiation over a wide range of wavelengths.

Figure 3.3. Eppley PSP pyranometer [Photo used with permission from Eppley Laboratories, Inc.].

3.2.2 Direct Sunlight Instruments

As their name implies, these instruments are designed to view only light coming directly from the sun. The radiation incident on one or more detectors is restricted to a narrow cone of the sky, the instrument's field of view. The field of view should be as small as possible in order to restrict the amount of scattered light that finds its way to the detector.

Direct sunlight instruments are called sun photometers or, in the case of broadband solar monitors, normal incidence pyrheliometers. Figure 3.4 shows a sun photometer manufactured by CIMEL Electronique and installed at La Parguera, Puerto Rico. The collimating tubes, which limit the field of view, and the detector housing below them, are together a little less than half a meter long. This instrument has

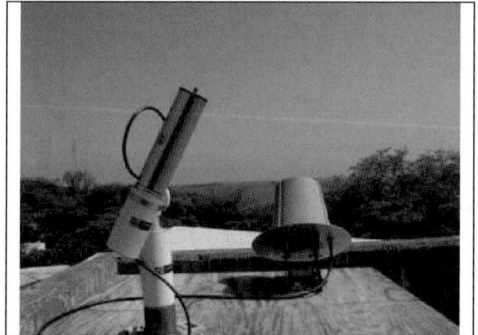

Figure 3.4. CIMEL sun photometer [http: //aeronet.gsfc.nasa.gov/new_web/photo_db/ La_Parguera.html].

its own onboard computer that is programmed to track the sun, and it automatically reports its data to a satellite receiver. (The hat-shaped object in the background is the antenna for communicating with the satellite.) The entire system operates on its own self-contained solar power system. These sun photometers are used by the Aerosol Robotic Network (AERONET), developed and managed by NASA's Goddard Space Flight Center, at sites around the world [Holben *et al.*, 1998].

3.2.3 Broadband Detectors

Broadband detectors are required for measuring total solar radiation. This is not as easy as it might seem, because solar radiation covers such a broad range of the electromagnetic spectrum. (Recall Figure 2.1.)

High-quality pyranometers for research that requires the most accurate measurements, such as the pyranometer shown in Figure 3.3, use thermopile detectors, made from collections of thermocouples. Thermocouples consist of dissimilar metals mechanically joined together. They produce a small voltage proportional to the temperature difference between their two components. When thermopiles are appropriately arranged and coated with a dull black finish, they serve as nearly perfect "blackbody" detectors that absorb energy across the entire range of the solar spectrum. These instruments are very expensive (several thousand dollars) and not practical for routine field work.

Photovoltaic detectors provide an alternative to thermopile detectors. Silicon-based solar cells are an obvious choice. Their major disadvantage is that their spectral response is different from the solar spectrum. Typically, they respond to sunlight in the range from roughly 400 to 1,100 nm, with a peak response in the near-infrared, around 900 nm. This restricted spectral response captures a subset of the solar spectrum which, under normal outdoor sunlight conditions, produces an insolation that is a few percent less than the total insolation. However, reliable and inexpensive solar radiation monitors are so desirable that a great deal of research effort has been dedicated to designing and understanding silicon-based pyranometers (*e.g.*, King and Myers [1997]). These instruments—they might more accurately be referred to as "surrogate pyranometers"—are *much* less expensive (a few hundred dollars for commercial products) than thermopile-based pyranometers. As will be shown in Chapter 4, it is possible to build a solar cell-based pyranometer for much less than the cost of commercial instruments (on the order of US$10 in parts). Such a pyranometer may be less rugged, or have a different cosine response than a commercial silicon-based instrument, but it is identical in principle.

3.2.4 Spectrally Selective Detectors

For both sun photometers and full-sky instruments, some applications require detectors that respond only to a specific range of wavelengths. For research instruments, the first choices for spectrally selective detectors are broadband photodetectors used in combination with so-called interference filters that transmit only a limited range of wavelengths, often only a few nanometers. These detectors are expensive, and they can be fragile and subject to unpredictable degradation. Thus, they are not good choices for student instruments. Detectors for the shorter UV wavelengths and for IR wavelengths beyond 2,000 nm or so tend to be very expensive, with no inexpensive alternatives. These realities impose some practical limitations on the kinds of spectrally selective measurements that can be made with inexpensive instruments. Nonetheless, a great deal of useful and interesting atmospheric science can be done with instruments that measure sunlight within the range of roughly 350–1,000 nm—the upper part of the UV through the visible and near-IR parts of the solar spectrum.

The first atmosphere monitoring instrument developed by the author and his colleagues for the GLOBE Program in the late 1990s, shown in Figure 3.5, uses light emitting diodes (LEDs) as detectors of green and red light [Brooks and Mims, 2001], based on the original concept by Mims [1992]. A near-infrared version for measuring total column water vapor, physically identical except for the detectors, was developed later [Brooks et al., 2003a].[1]

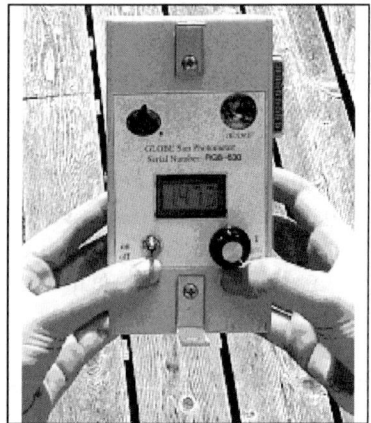

Figure 3.5. Two-channel LED-based sun photometer.

The physical laws that cause LEDs to emit light when a current is passed through them also work the other way around: LEDs generate a small electrical current when light in an appropriate wavelength range shines on them. LEDs are inexpensive, stable, and virtually indestructible— essential attributes for instruments intended for student use.

Of course, LEDs have some disadvantages. The main difficulty is that even though LEDs can serve as spectrally selective light detectors,

---

[1] Some versions of the water vapor instrument used one LED and one filtered photodiode. Other versions used two filtered photodiodes.

they are designed to *emit* light, not *detect* it, and they are optimized for their intended purpose. The spectral response of LEDs is often wider than is desirable for some kinds of atmospheric measurements, and the available wavelength responses may not be ideal for the intended measurement. The spectral response of an LED is not simply related to its emission spectrum. The current generated by any photodetector exposed to light varies with temperature and this fact can limit the usefulness of LED detectors if there is a strong temperature dependence. However, despite these potential problems, there are still many scientifically valid applications of LED detectors in the UV, visible, and near-IR.

3.3 What Can You Measure?

Many interesting measurements can be made with the kinds of full-sky and direct sunlight instruments discussed above. These measurements can have scientific as well as educational value. For example, consider validating measurements made from Earth orbit. The satellite-based instrument has a particular resolution on the ground, called its "footprint." Sometimes, this footprint is tens or even hundreds of kilometers on a side. Thus, a single measurement from space represents an instantaneous spatial average over distances that may be, for example, large compared to typical cloud dimensions—essentially, a fuzzy "snapshot" in both space and time that renders it impossible to separate cloud and ground contributions to the space-based measurement.

How can a "fuzzy" space-based measurement be related to what an observer sees looking up toward the satellite from the ground? (This is an especially difficult question under partly cloudy skies.) What are the sky conditions before and after a satellite instrument has moved on along its orbit? The answers to these questions may require several instruments distributed around a satellite instrument footprint, collecting data continuously. With research-quality instruments, such a spatially dense instrumentation network can be prohibitively expensive. However, one reference instrument used in conjunction with several very inexpensive instruments can make such an undertaking possible.

3.3.1 Total Solar Radiation

The total amount of solar radiation reaching a horizontal plane at Earth's surface—the insolation, measured in units of W/m^2—is a fundamental measurement for understanding Earth and its atmosphere as an interconnected dynamic system. There are obvious engineering applications of such a measurement, for designing and siting solar power systems. From a science perspective, long-term solar monitoring helps scientists understand persistent changes in the atmosphere.

The solar radiation at the top of the atmosphere over a particular site is determined entirely by the time of year and latitude. However, interaction with the atmosphere and clouds alters this simple geometric distribution of solar energy reaching Earth's surface. Around noon on a clear summer day in temperate climates, the insolation at the surface is roughly 1,000 W/m². During the winter, the insolation can be less than half that amount. Figure 3.6 shows a false-color image of modeled 4-day mean insolation over North America in late December, 2003, based on analysis of GOES visible images [Diak *et al.*, 1996]. The units are megajoules per day. The color bands corresponding to different levels of insolation clearly do not follow lines of constant latitude, so the effects of cloud cover are clear.

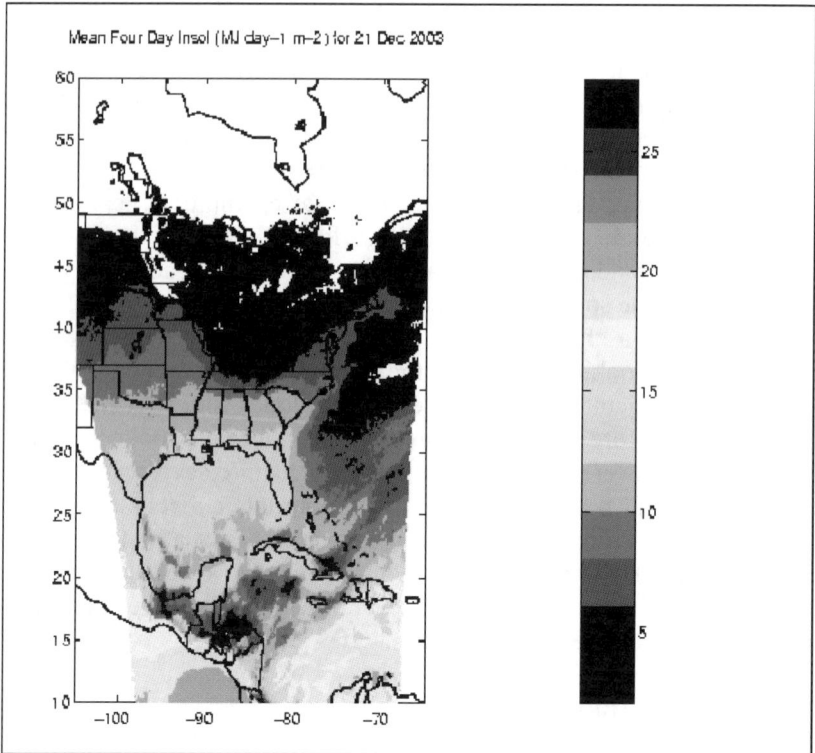

Figure 3.6 (see color plates). 4-day mean insolation over North America, late December 2003, based on GOES visible images [Diak *et al.*, 1996].

There are also less obvious applications of insolation data. For example, within a theoretical envelope provided by the diurnal solar cycle under clear skies at a specified time of year, site latitude, and elevation, it is possible to relate temporal and spatial fluctuations of insolation to air quality and cloud cover. So, not only are the average effects of cloud cover embedded in pyranometer data, it is even possible to build cloud statistics to form a local cloud climatology [Duchon and O'Malley, 1999]. This is valuable because the analysis methods are completely objective and do not rely on more subjective and less diligent human observers.

These possibilities can be seen in Figure 3.7. These data are taken from a 2-year set of pyranometer data recorded at 1-min intervals at Central Middle School, Waterloo, Iowa. The values shown in boxes over each day's data are the maximum observed insolation and the daily average insolation. In these data, the effects of clouds moving across the observing site are clearly evident, including the phenomenon by which sunlight reflected from the sides of clouds causes temporary large "spikes" in insolation reaching the surface. On June 12, for example, the data show clouds moving across the sun at this observing site around noon, following a clear morning. Note, however, that momentarily large insolation values from cloud reflections are always more than balanced out by low values. Of the days shown in Figure 3.7, the clearest day, June 16, has the highest daily average insolation.

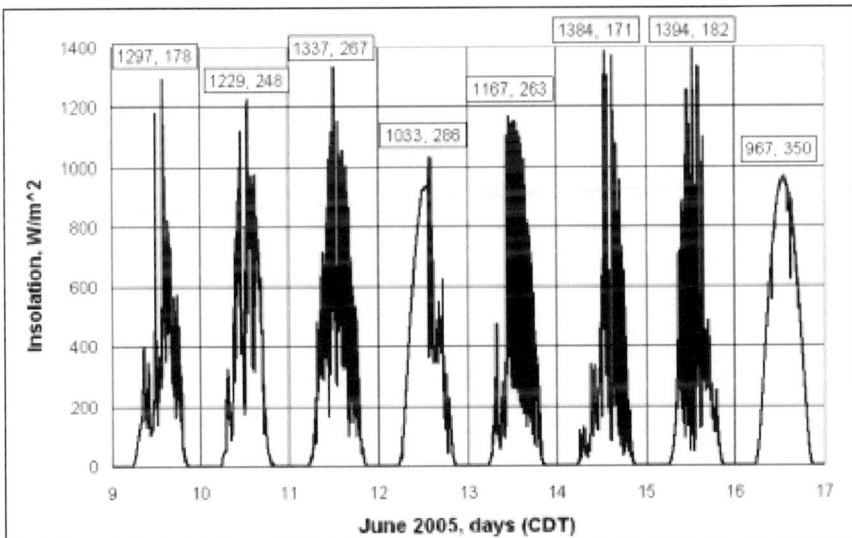

Figure 3.7. Insolation measurements, June 2005, Central Middle School, Waterloo, Iowa.

3.3.2 Aerosols

Molecules and liquid or solid particles suspended in the atmosphere (aerosols) scatter and absorb sunlight. The effect of molecular scattering on direct sunlight can be calculated theoretically as a function of wavelength, as can the effect of absorption by gases. The remaining reduction in direct sunlight at a particular wavelength is due to scattering and absorption by aerosols in the atmosphere. These reductions are described by a quantity known as optical thickness, or optical depth. The more scattering and absorption reduce transmitted sunlight as viewed along a direct path, the larger the optical thickness. The basic equation governing the transmission of radiation through an intervening medium is the Lambert/Beer/Bouaguer law, often called simply Beer's law):

$$I_\lambda = I_{o,\lambda} \exp(-\alpha_\lambda m_{air}) \tag{3.2}$$

where $I_{o,\lambda}$ is the original source intensity, I_λ is the intensity after radiation passes through a relative air mass m_{air}, and α_λ is the total atmospheric optical thickness (including molecular scattering and scattering and absorption by gases and aerosols), all at wavelength λ. The reference to wavelength is important because Beer's law applies, in principle, only to a specific wavelength of light.

Relative air mass is a dimensionless quantity equal to 1 when the sun is directly overhead at a zenith angle of 0° and approximately equal to the inverse of the cosine of the solar zenith angle:

$$m_{air} \approx 1/\cos(z) \tag{3.3}$$

A more accurate formulation is given in Appendix 4.

Perhaps a more intuitive formulation of optical thickness describes the same effect in terms of percent transmission through the atmosphere T:

$$T = 100 \cdot \exp(-\alpha_\lambda) \tag{3.4}$$

The relationship between percent transmission and optical thickness is shown in Figure 3.8. The larger the optical thickness, the less direct sunlight is transmitted. A typical value of optical thickness for visible light in clean air near sea level is around 0.2, or 81.9% transmission.

The total optical thickness can be separated into its three components, corresponding to molecular (Rayleigh) scattering, absorption by gases such as ozone, and scattering and/or absorption by aerosols:

$$\alpha_\lambda = \alpha_{\lambda,R} + \alpha_{\lambda,g} + \alpha_{\lambda,a} \tag{3.5}$$

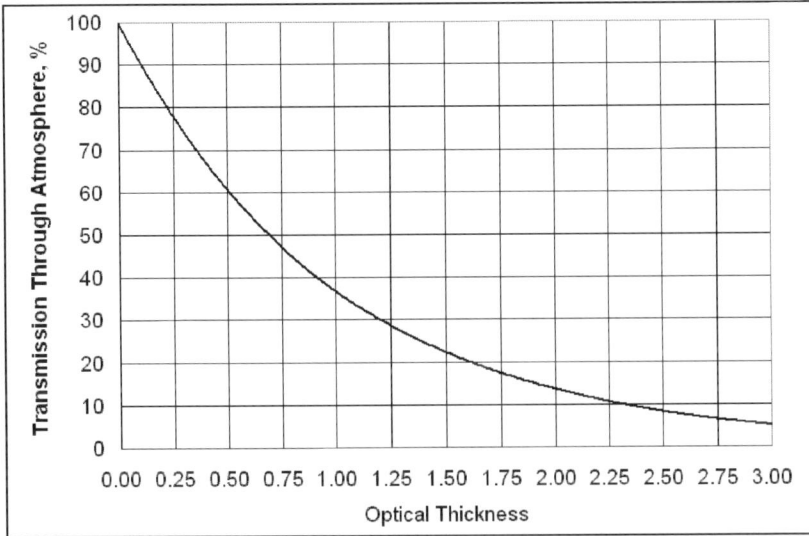

Figure 3.8. Percent transmission through the atmosphere vs. total optical thickness.

Each of these terms is wavelength-dependent, which explains why optical thickness values are associated with a particular wavelength. The strongly wavelength-dependent contribution of Rayleigh scattering can be calculated using theoretical models of the atmosphere. For practical calculations, a parameterized model is used. One formulation has been given by Bucholtz [1995]:

$$\alpha_{\lambda,R} = (p/p_0)A \cdot \lambda^{(-B - C\lambda - D/\lambda)} \tag{3.6}$$

where (p/p_0) is the ratio of actual barometric pressure at an observing site to standard pressure at sea level (1013.25 mbar) and λ is the wavelength in units of microns. The parameters A, B, C, and D given in Table 3.1 are empirically derived values that provide a mathematical best fit to theoretical calculations. Figure 3.9 shows the Rayleigh coefficient at sea level as a function of wavelength.

Table 3.1. Coefficients for calculating Rayleigh optical thickness [Bucholtz, 1995].

Coefficient	$\lambda \leq 0.500 \ \mu m$ (500 nm)	$\lambda > 0.500 \ \mu m$ (500 nm)
A	6.50362×10^{-3}	8.64627×10^{-3}
B	3.55212	3.99668
C	1.35579	1.10298×10^{-2}
D	0.11563	2.71393×10^{-2}

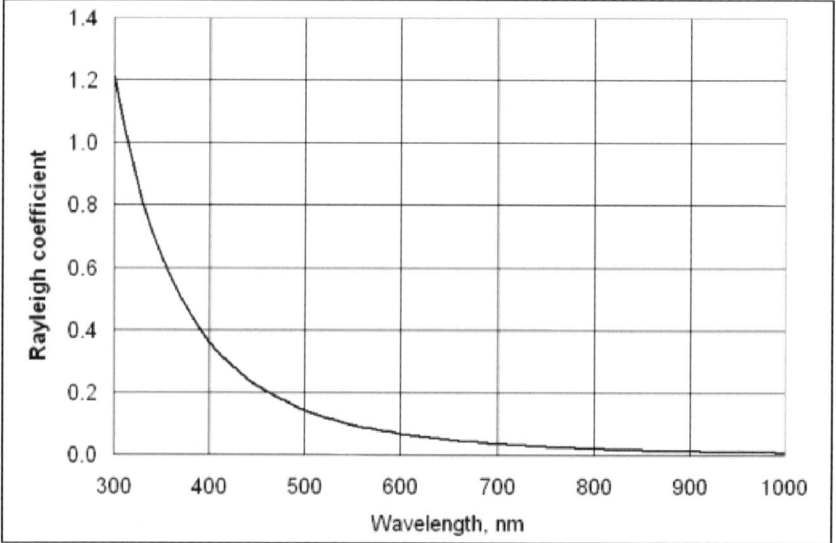

Figure 3.9. Dependence of Rayleigh scattering coefficient on wavelength at standard atmospheric pressure.

The predominant gas absorber over the range of visible wavelengths is ozone. Figure 3.10 shows the climatological mean optical thickness for ozone across visible and near-IR wavelengths [Leckner, 1978; Bird and Riordan, 1986]. Although it is small (<0.04), it can be significant for clean air with small values of aerosol optical thickness.

The remaining contribution to optical thickness comes from aerosols. Sources of aerosols include volcanic activity, wind-blown dust from deserts and agricultural activity, sea spray, and air pollution. The

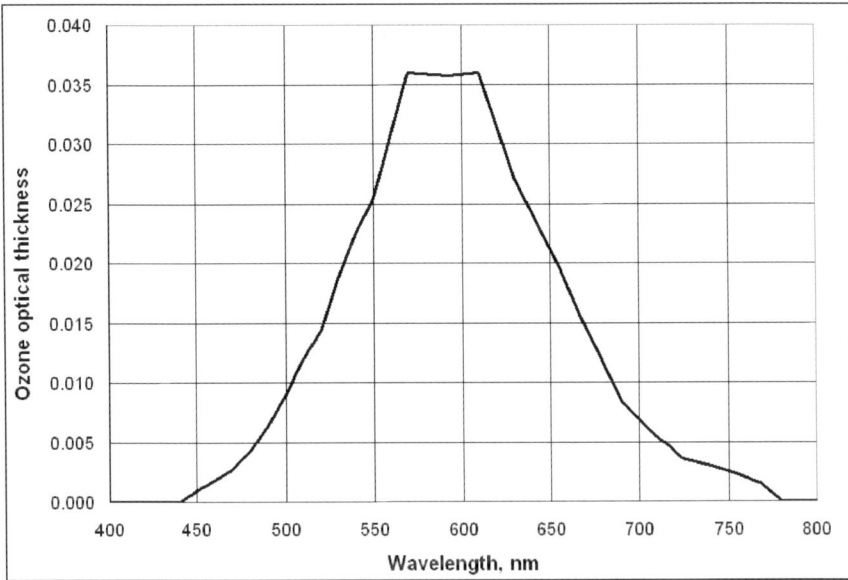

Figure 3.10. Ozone optical thickness at visible and near-IR wavelengths.

amount of aerosols varies widely around the globe, and there are strong seasonal effects. In clean skies, aerosol optical thickness (AOT) at visible wavelengths will be less than 0.1.

Figure 3.11 shows a false-color image representing the monthly mean of global aerosol optical thickness for July 2006, generated from data recorded by the Moderate Resolution Imaging Spectroradiometer (MODIS) instrument orbiting Earth on NASA's EOS/Terra spacecraft [NASA, 2008]. Cloudy weather and/or snow cover over the northern hemisphere prevents AOT calculations from being done in some places. Bright land surfaces, such as the Sahara Desert, the Saudi Arabian Peninsula, and permanently snow/ice-covered polar regions, are inaccessible to MODIS aerosol retrievals. Thus, ground-based measurements are still very important.

High values of AOT resulting from dust and biomass burning activity are clearly evident in the sub-Sahara, southern Africa, and the Indian subcontinent. A large plume of dust and smoke is blowing westward from Africa toward the Caribbean. Large amounts of smoke and dust are visible over Asia. These aerosols often travel eastward across the Pacific Ocean to the western United States.

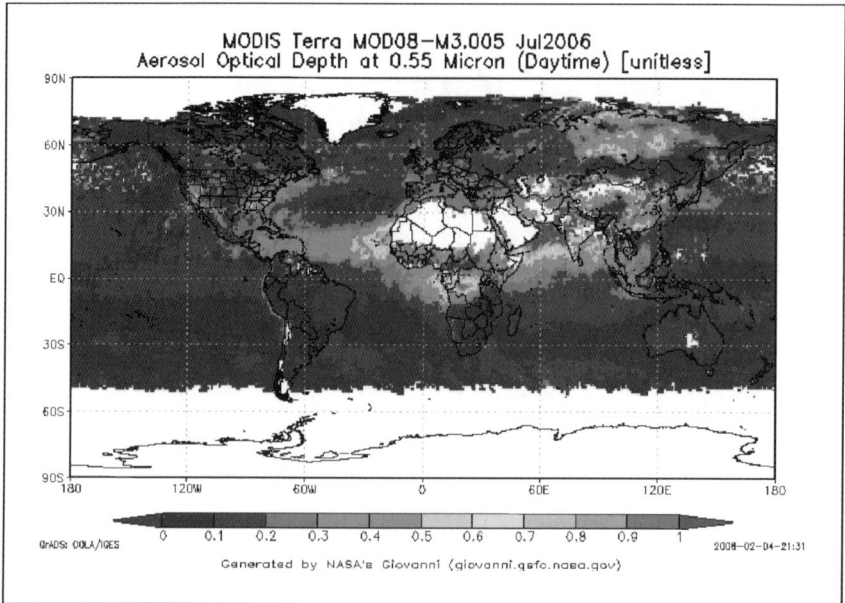

Figure 3.11 (see color plates). 550-nm aerosol optical depth from MODIS/Terra, monthly mean values for July, 2006.

3.3.3 Photosynthetically Active Radiation

Photosynthetically active radiation (PAR), as its name implies, refers to the subset of total solar radiation that plants use for photosynthesis. Typically, this is a full-sky measurement. The spectral definition of PAR is not precise, because different kinds of plants respond to different parts of the solar spectrum. However, PAR is generally considered to include only the visible part of the solar spectrum—between 400 and 700 nm. Thus, a PAR detector is similar to a pyranometer, but with its spectral response limited to visible wavelengths. This is an extremely important measurement in agriculture and botany for studying plant behavior under different lighting conditions—under crop or tree canopies, for example.

Although it is true that the output from a full-sky PAR detector is strongly correlated with total solar radiation, scientists require a separate measurement for PAR radiation. Whereas insolation is reported in units of energy density (W/m^2), PAR radiation is reported in terms of the total number of photons in the range between 400 and 700 nm. This makes sense because photosynthesis involves the interaction between individual photons and molecules. Specifically, PAR is reported as the number of moles (usually micro-moles) of photons per unit area per second in the spectral interval from 400 to 700 nm, μ-moles/m^2/s, where there are

6.0222×10^{23} (Avogadro's number) photons per mole. Figure 3.12 shows PAR under partly cloudy skies at the USDA UV-B Monitoring Network at Beltsville, Maryland, on 1 September, 2006. These data look similar to insolation for this day, in their response to moving cloud patterns, but, of course, the quantity being measured is different.

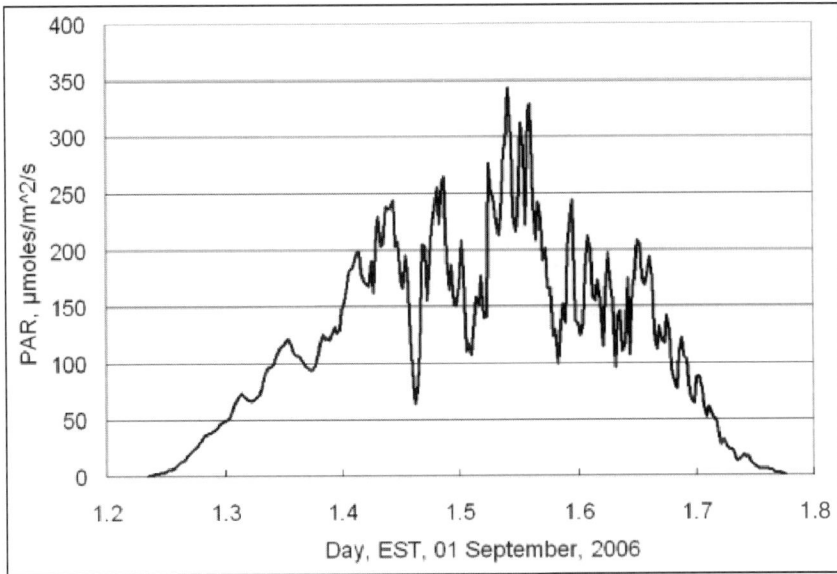

Figure 3.12. Photosynthetically active radiation (PAR) at Beltsville, Maryland, 1 September 2006.

3.3.4 Water Vapor

Total atmospheric water vapor, also called total column water vapor or total precipitable water vapor (PW) is defined as the thickness of a layer of water obtained by condensing all the water vapor in a column above the observer and bringing it down to the observer's elevation. Typically, PW is a few centimeters.

PW is distributed very unevenly around the globe. Figure 3.13 shows a view of the Western hemisphere on 22 December 2003, based on infrared (6,750 nm) data from the GOES-12 satellite. The lighter the color, the more water vapor there is in the atmosphere. In general, there is more water vapor over warm water and equatorial forests because of evaporation and transpiration. Not surprisingly, there is much less water vapor over deserts and at high elevations. Although it is commonly believed that there must be a lot of precipitation at the poles, because they

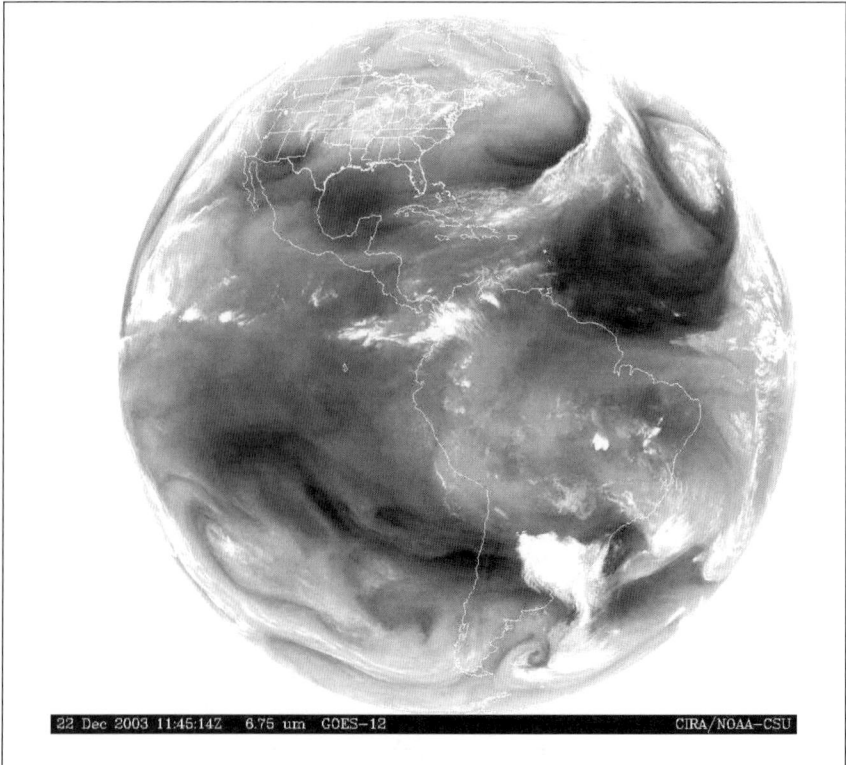

Figure 3.13. GOES-12 6,750 nm water vapor image for the Western hemisphere, 11:45 UT, 22 December, 2003 [See http://www.ncdc.noaa.gov/oa/satellite.html].

are covered with snow and ice, the air over Earth's polar regions is often very dry.

At a given location, PW varies seasonally and diurnally. Long-term global changes in PW can signal climate change. Figure 3.14 shows a 12-year record of PW over Seguin, Texas, USA [Mims, 2002]. This figure illustrates some effects associated with major volcanic eruptions such as Mt. Pinatubo in 1991 (higher levels of PW in the following winter) and the strong El Niño event around 1997 (higher levels of PW).

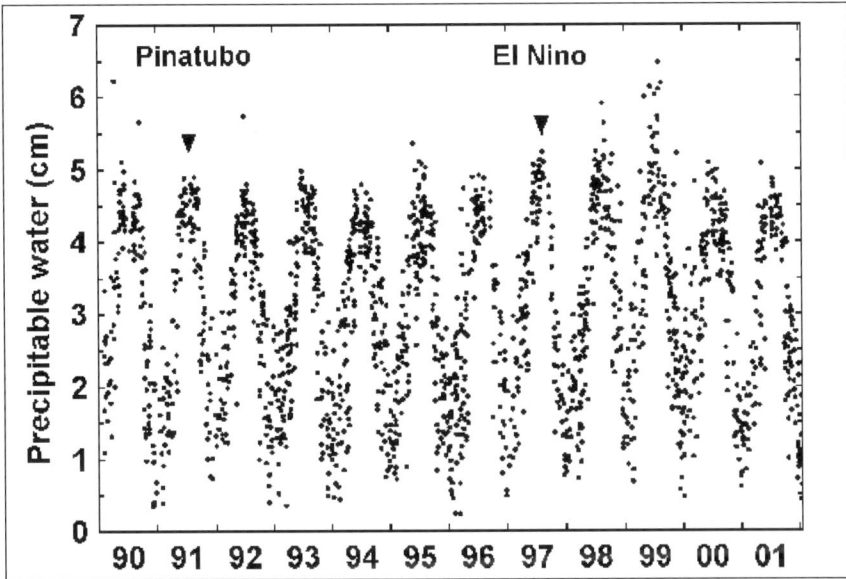

Figure 3.14. Precipitable water over Seguin, Texas, USA [Mims, 2002].

3.3.5 Ultraviolet Radiation

As described in Chapter 2, radiation from the sun covers a wide range of the electromagnetic spectrum. Ultraviolet radiation, with wavelengths just below the visible spectrum, is especially important to life on Earth. The UV spectrum is usually defined as radiation with wavelengths between 100 and 400 nm. There are three widely used UV categories, as shown in Table 3.2.

Table 3.2. UV categories, by wavelength.

UV category	Wavelength range (nm)
UV-A	315–400
UV-B	280–315
UV-C	100–280

Figure 3.15 shows 317-nm and 368-nm radiation measured with an Ultraviolet Multi-Filter Rotating Shadowband Radiometer manufactured by Yankee Environmental Systems, at the USDA research site in Beltsville, Maryland, USA. These wavelengths are in the UV-A band as indicated in Table 3.2.

UV-C radiation can damage DNA. It will kill bacteria and viruses. In fact, artificial UV-C sources are used to sterilize medical equipment and to purify air and water. There is virtually no naturally occurring UV-C radiation at Earth's surface because it is absorbed by oxygen in the atmosphere. The interaction of UV-C radiation with oxygen is the source of stratospheric ozone.

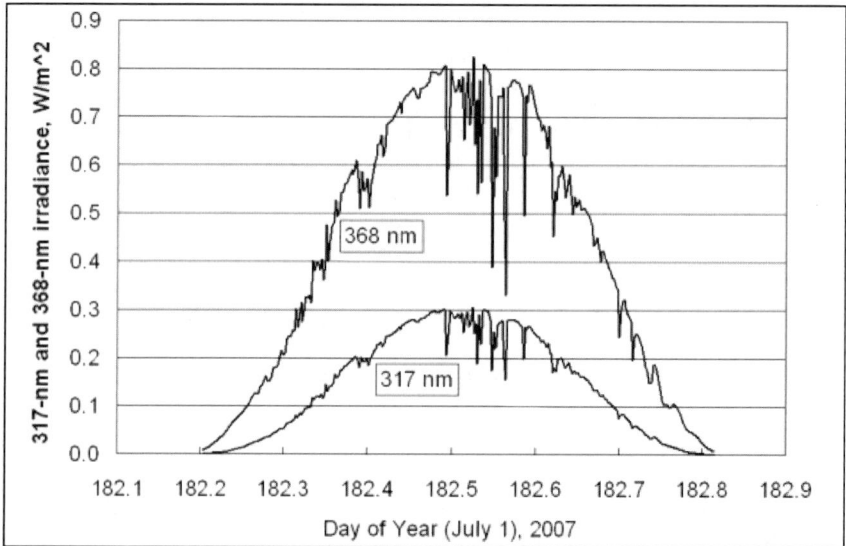

Figure 3.15. 317-nm and 368-nm UV irradiance at Beltsville, Maryland, July 1, 2007.

UV-B is partially absorbed by ozone in the stratosphere. It is considered a destructive form of UV radiation at Earth's surface because it can damage living tissue and manmade materials. Increased human exposure to UV-B radiation from sunlight produces sunburn and is widely accepted as the cause of increasing rates of the most serious forms of skin cancer (melanomas). Even though overexposure to the sun is a serious human health problem, "getting a tan" is still considered important by many light-skinned individuals. Because UV-B is only partially absorbed by ozone, even a small decrease in the amount of stratospheric ozone can significantly increase the risk of skin cancer for light-skinned humans. Overexposure to UV-B is also associated with eye diseases such as cataracts, which can affect humans regardless of their skin color.

Low-energy UV-A ("black light") sources can cause some materials to emit visible light (fluoresce), which makes them appear to glow in the dark. UV-A radiation penetrates farther into human skin than

UV-B. Tanning lamps are designed to produce UV-A rather than UV-B radiation because exposure to UV-A will darken light skin without burning. However, UV-A exposure can cause premature skin aging and eye problems. Therefore, health professionals warn against excessive UV exposure, including UV-A, no matter what the source.

Because of the direct connection between stratospheric ozone and the amount of UV radiation reaching Earth's surface, measurements of UV radiation are of great scientific interest. At ground level, UV has some important ecological effects. For examples, many insects can see UV light and disruption in the UV environment within which they have evolved can affect their ability to find food and breed.

Although UV radiation is of special concern because over-exposure is a serious health issue for humans, it is important to understand that the interactions of UV radiation with life on Earth are complex and, in many cases, poorly understood. UV radiation is inherently neither "good" nor "bad." Life on Earth has evolved within a particular radiation environment that includes some UV radiation. Disruptions to this radiation environment, causing some forms of radiation to increase or decrease, can have serious and unforeseen consequences. This is the source of concern about "ozone holes" in the stratosphere.

3.3.6 Surface Reflectance

A space-based measurement of great scientific interest is surface reflect-ance. Especially when made in several spectral bands, measurements of surface reflectance can be used to detect seasonal changes ("green-up" and "green-down" of vegetation), land use changes, and the health of vege-tation. At Earth's surface, reflectance measurements are also interesting. Figure 3.16 shows a false-color representation of Earth's broadband albedo (reflectance) averaged over the month of October 1986. It is based on assumptions about the underlying surface types and their properties, plus space-based measurements of ice and snow. Ground-based measurements are important for validating the quality of such maps, and for monitoring changes in surface properties as a function of time.

Any of the full-sky instruments described in this book can be used in pairs to make reflectance measurements over a variety of surfaces. One advantage of these measurements is that they do not require an absolute instrument calibration. Reflectance is defined as the ratio of responses from an upward-looking instrument and an identical downward-looking instrument.

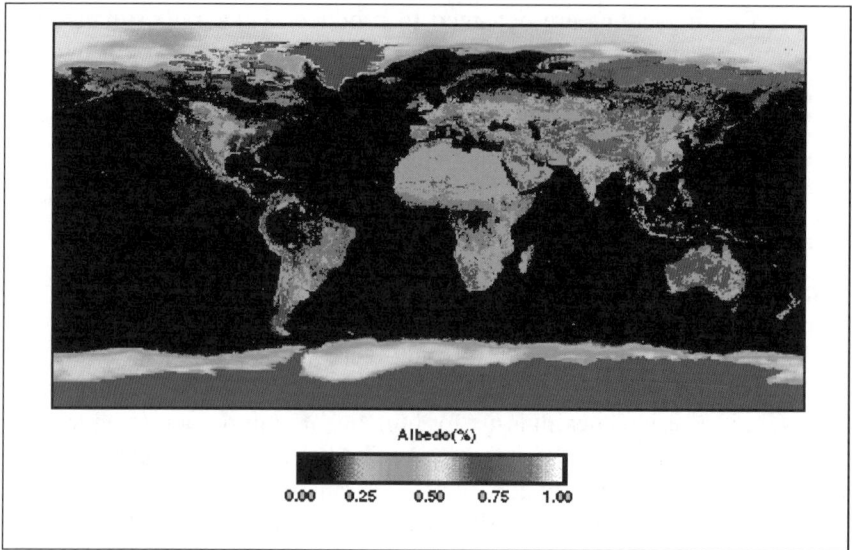

Figure 3.16 (see color plates). False-color representation of monthly average broad-band albedo, October, 1986 [www-surf.larc.nasa.gov/surf/pages/bbalb.html].

Even "identical" instruments will not, in general, produce exactly the same output, but this calibration problem is easily solved simply by defining one instrument as the "reference" and calibrating the other instrument to agree with this reference when they are side by side and pointing in the same direction.

3.4 Making the Transition from Ideas to Measurements

This chapter has presented some scientifically interesting measurements related to solar transmission through the atmosphere. These examples rely on space-based measurements or ground-based equipment that is far too expensive for this book's intended audience. The challenge is to develop methods of doing equally interesting science with instruments that are both reliable and inexpensive. This is not a trivial undertaking, but it is very instructive because the underlying design process is the same regardless of whether an instrument costs $20 or $200,000 (or more!), and whether it is based on the ground or on an orbiting platform. Questions that must always be asked include:

1. Do I understand the physical principles on which the measurement is based?
2. Do I understand how to translate those principles into a reliable and practical measuring system?
3. Do I understand the potential and the limitations of the instrument?
4. Do I understand how to calibrate and maintain the instrument?
5. Do I understand how to record data from the instrument and how the recording process influences the performance of the instrument?
6. Do I understand how to interpret output from the instrument?
7. Do I understand how to relate my measurements to others?

Subsequent chapters will address these questions for several instruments, starting in Chapter 4 with an inexpensive pyranometer.

4. Instrument Design Principles I: Radiometers

Chapter 4 examines the design of instruments for measuring the total amount of radiation, or spectrally restricted subsets of the total radiation, reaching or reflected from Earth's surface. A detailed discussion of the design of pyranometers is given, to illustrate how to transform design principles into working instruments. An introduction is given to using light-emitting diodes as spectrally selective sunlight detectors. For each instrument, calibration procedures and applications are discussed.

4.1 Measuring Total Solar Radiation

4.1.1 Choosing and Characterizing a Broadband Detector for a Pyranometer

As described in Chapter 3, an instrument that measures total solar energy, or insolation, is called a pyranometer. There are basically two kinds of pyranometers. The expensive kind (several thousand dollars) uses thermo-pile detectors—collections of thermocouples—configured in a circular pattern under a high-quality glass or quartz dome. (Recall Figure 3.3.) These instruments are often considered too expensive even for "professional" use as routine insolation monitors.

A much less expensive alternative is to use silicon solar cell detectors. Silicon solar cells are semiconductor devices consisting of thin sheets of silicon "doped" with impurities. Two dopants are used—one has an excess of unbound electrons (an N-type dopant) and the other has a deficit of electrons (a P-type dopant). This doping creates an electric field. When light strikes the cell, the energy from individual photons is absorbed by atoms and creates a potential difference—the "open-circuit" voltage. If the positive and negative terminals are connected, electrons flow and generate a "short-circuit" current, the size of which depends on the properties of the solar cell.

A basic silicon-cell pyranometer (SIP) is an extremely simple device. Figure 4.1 shows the circuit schematic, which consists of a solar cell and a current-measuring device. When exposed to light, the solar cell generates a flow of electrons which produce an electrical current, I. The amount of current is linearly proportional to the solar energy incident on the cell.

The major problem with SIPs is that they respond only to a subset of the total solar spectrum, and this response is very uneven even across the wavelengths in that subset. The spectral response of a typical solar-cell-based pyranometer is shown in Figure 4.2.

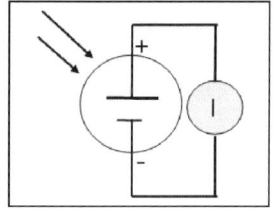

Figure 4.1. Circuit diagram for a simple solar-cell-based pyranometer.

This response is so uneven and restricted that it might be more accurate to call these instruments "surrogate pyranometers." Nonetheless, despite these problems, SIPs are widely used for meteorological, agricultural, and environmental monitoring, and their performance relative to thermopile-based pyranometers has been studied extensively for decades.

In order to transform the schematic in Figure 4.1 into a practical, working pyranometer, it is first necessary to understand more about how solar cells work. A solar cell is characterized by three parameters: its open circuit voltage, its short circuit current, and its ability to do "work," in the physics sense.

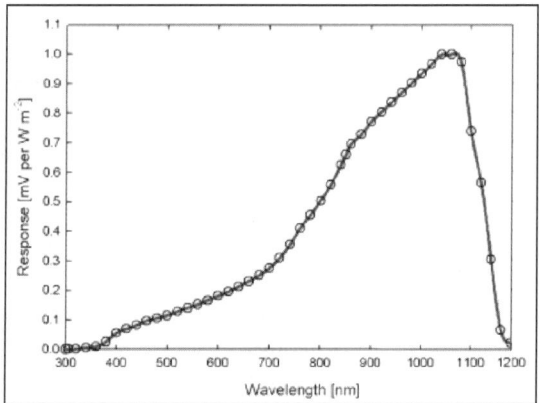

Figure 4.2. Normalized spectral response of a silicon solar-cell pyranometer [Courtesy of Apogee Instruments, Inc., 2008].

Figure 4.3 shows measurements made with a solar cell purchased from a surplus electronics company. The cell has solder pads conveniently located on the back, to which red ("+") and black ("–") wires are attached. A 100-Ω resistor is attached to the end of the wire from the "–" terminal. All measurements are made in full sunlight. The open-circuit voltage of this solar cell (4.26 V, as shown in the top image) is defined as the voltage generated when no electrons are flowing, measured directly across the "+" and "–" terminals (not through the resistor). The short-circuit current (82.3 mA, as shown in the middle image) is the current flowing directly between the "+" and "–" terminals with no resistance in the circuit.

Solar cells have a power rating that is determined by their ability to do work across a load—a resistance that could be a light bulb, a small motor, or in this case simply a 100-Ω resistor. With the resistor attached to the "+" terminal and connected to the "–" terminal through a multimeter set to measure current (the multimeter on the left in the bottom image), this cell produces 39.5 mA. The multimeter on the right measures voltage across the resistor, 3.88 V. The power P generated by this cell is current times voltage (or equivalently, as shown in Figure 4.3, V^2/R, using Ohm's Law, $V = I \cdot R$).

$$P = \text{current} \times \text{voltage} = I \cdot V = 0.0395 \cdot 3.88 = 0.15 \text{ W} \qquad (4.1)$$

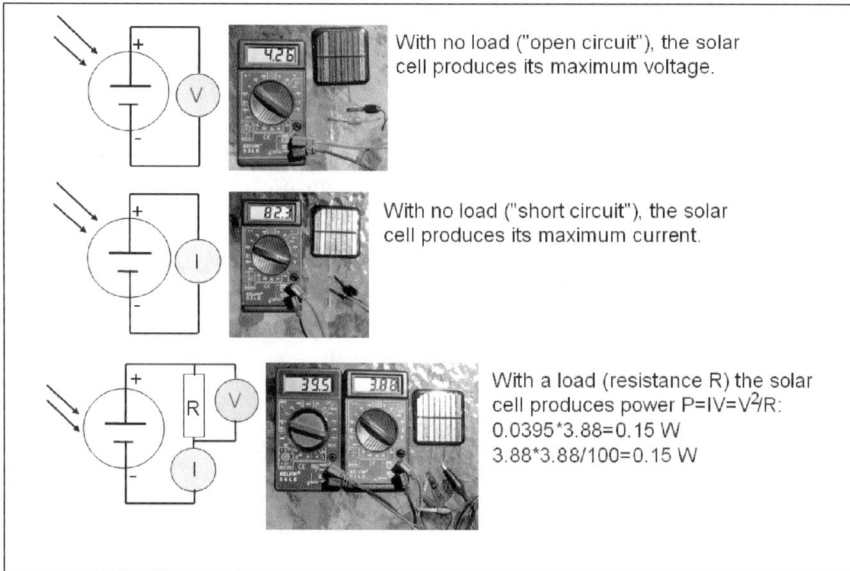

With no load ("open circuit"), the solar cell produces its maximum voltage.

With no load ("short circuit"), the solar cell produces its maximum current.

With a load (resistance R) the solar cell produces power $P=IV=V^2/R$:
0.0395*3.88=0.15 W
3.88*3.88/100=0.15 W

Figure 4.3 (see color plates). Measurements on a solar cell: open-circuit voltage, short-circuit current, and work across a resistor.

Recall from Chapter 3 that around noon on a clear summer day in temperate climates, about 1,000 W/m^2 of solar energy reaches Earth's surface. The solar cell shown in Figure 4.3 measures about 6 × 6 cm. Thus a packed 16 × 16 array of these solar cells (about 1 m^2) would produce only about 256·0.15 = 38.4 W/m^2 with a 100-Ω load, for a conversion efficiency of a little less than 4%.

This particular cell is certainly not state-of-the-art. It is a surplus item, after all, and is clearly designed more for durability and ease of use than for optimum power production. Probably the 100-Ω load does not allow this solar cell to produce its maximum power output. Regardless of the limitation of this particular solar cell, the measurements described in Figure 4.3 demonstrate that the direct conversion of solar energy to electricity is not a very inefficient process.

It is worth thinking carefully about assumptions that are made when current and voltage are measured as described in Figure 4.3. When measuring the open-circuit voltage, the assumption is that no electrons are flowing. When measuring the short-circuit current, the assumption is that there is no resistance to the flow of electrons. However, a measuring instrument must divert a few electrons to determine the voltage produced by a circuit, and it must provide a small resistance to the flow of electrons to determine current flowing through a circuit. That is, these measurements can be made only by violating, even if in a very small way, the assumptions that have been made.

This point is important for designing a pyranometer, because the assumption is that current output is linearly proportional to incident solar radiation. In order to record this output with a data logger, as will be discussed in more detail later, the output current must be converted to voltage in a way that maintains the linear relationship between current and incident solar radiation.

A secondary but also important consideration in pyranometer design is that the detector must have a good cosine response to a direct beam of sunlight, as previously described in Chapter 3. (Refer to the discussion of equation 3.1.) If a direct beam of sunlight falls on a horizontal surface, its intensity varies theoretically as the cosine of the zenith angle. You should not be surprised to learn that real detectors do not have a perfect cosine response! As a practical design matter, it is much easier to deal with this problem by using a

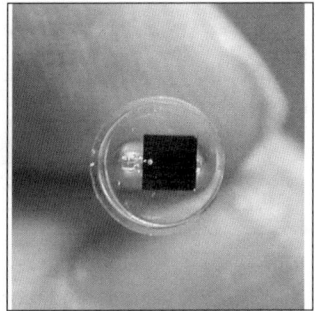

Figure 4.4. A PDB-C139 silicon photodiode.

small detector covered with a diffusing material, based on the assumption that the diffuser will greatly reduce some of the problems with reflections that a "bare" detector will have, thereby improving the cosine response.

In order to make additional progress on the design of a practical pyranometer, it is instructive to examine the performance of a suitable detector. Figure 4.4 shows an end view of a small blue-enhanced silicon photodiode (Advanced Photonix part PDB-C139) in a T 1-3/4 clear epoxy housing. This is one of the standard sizes for LED housings, measuring 5 mm in diameter. The solar cell itself has an active area of about 2×2 mm^2.

Figure 4.5 shows this detector mounted in a plastic holder machined from 1/4" (approximate inside diameter) "Schedule 80" thick-walled PVC plumbing pipe and covered with a 1-cm diameter, 1-mm-thick Teflon® diffusing disk. For demonstration purposes, an analog milliammeter is used to measure the current produced by this device. In today's "digital world," many students have never used analog meters, but they are a very useful reminder that the world is an analog place, and they force us to be aware of the physical basis of measurements. In this case the solar cell is forced to do a very small amount of work (equivalent to providing a small resistance to the flow of electrons) by generating a tiny magnetic field in a coil of wire which causes the meter's needle to rotate. It might seem that the needle is too large to be driven by such a tiny solar cell, but the entire electromechanical assembly is very delicately balanced. This cell produces what is still very nearly a short-circuit current of about 0.55 mA.

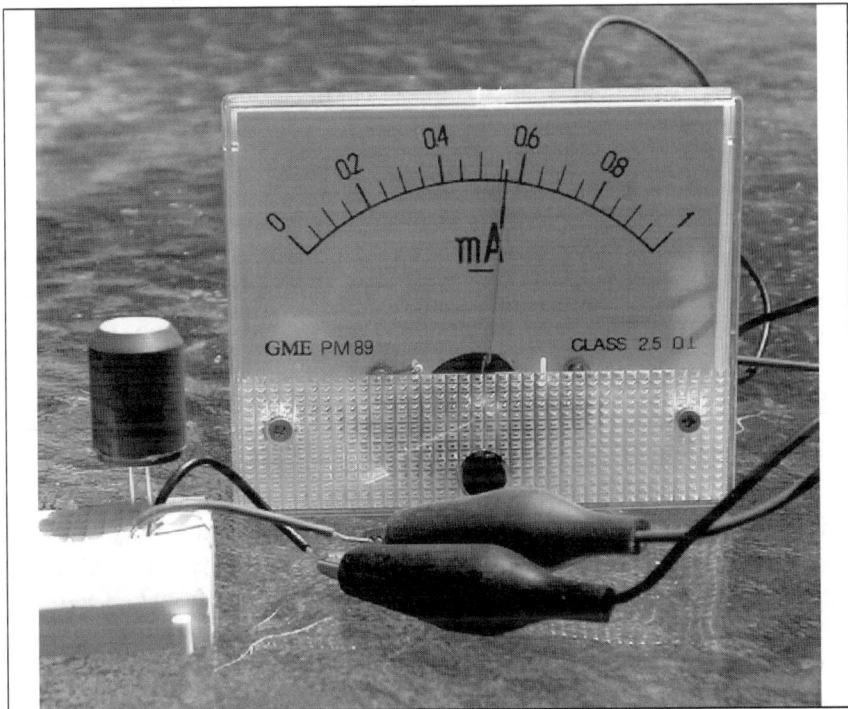

Figure 4.5 (see color plates) Current output from PDB C139 silicon photodiode with diffuser.

This detector produces an open-circuit voltage of a little more than 0.5 V but as already noted, this characteristic is not particularly relevant to the design of a pyranometer. This situation is illustrated in Figure 4.6, which shows the open-circuit voltage of the device shown in Figure 4.5 recorded next to the voltage produced by a commercial pyranometer (an Apogee PYR-P), starting around mid-morning on a clear summer day. The open-circuit voltage remains nearly constant until dark, but the "real" pyranometer voltage output is proportional to the incident solar energy, and describes, on average, a cosine-like curve as the sun sets in the sky. On this day, cumulus clouds moved into the observing area during late morning. The Apogee pyranometer responds to the reflection and shadowing of sunlight by these clouds, but the photodetector open-circuit voltage changes by a much smaller amount.

Figure 4.6 should make clear that the open circuit voltage from a photodetector should *not* be used to measure insolation. Actually, the only reason the "open-circuit" voltage in this example fluctuates as much as it does during the day is that the logger used to record the data does not have a very high input impedance, and therefore it does not record the true open-circuit voltage. This matter is discussed in Appendix 6.

Since it is clear that simply recording the open-circuit voltage of a detector is not appropriate, what *should* you record? To be consistent with the goals of designing inexpensive instruments, recording should require only an inexpensive data logger. But, as described in Appendix 6, data loggers are voltage-recording devices, and they cannot be interfaced directly with current-producing devices. This means that the output signal from a pyranometer needs to be large enough so that sufficient resolution can be obtained with an inexpensive data logger, but the signal must also remain linearly proportional to current—that is, linearly proportional to the incident solar energy.

The solution for a pyranometer is to force the solar-cell detector to do work across a load resistance, as previously described in Figure 4.3. This will develop a voltage, as demonstrated in Figure 4.6 with the Apogee pyranometer. But will this voltage be linearly proportional to the incident solar energy, in the same way as short-circuit current? Even if the voltage is linearly proportional to the incident solar energy, will the available output from this small detector be sufficient to drive an inexpensive data logger? These are not trivial questions, and answering them requires some additional testing of the proposed pyranometer detector.

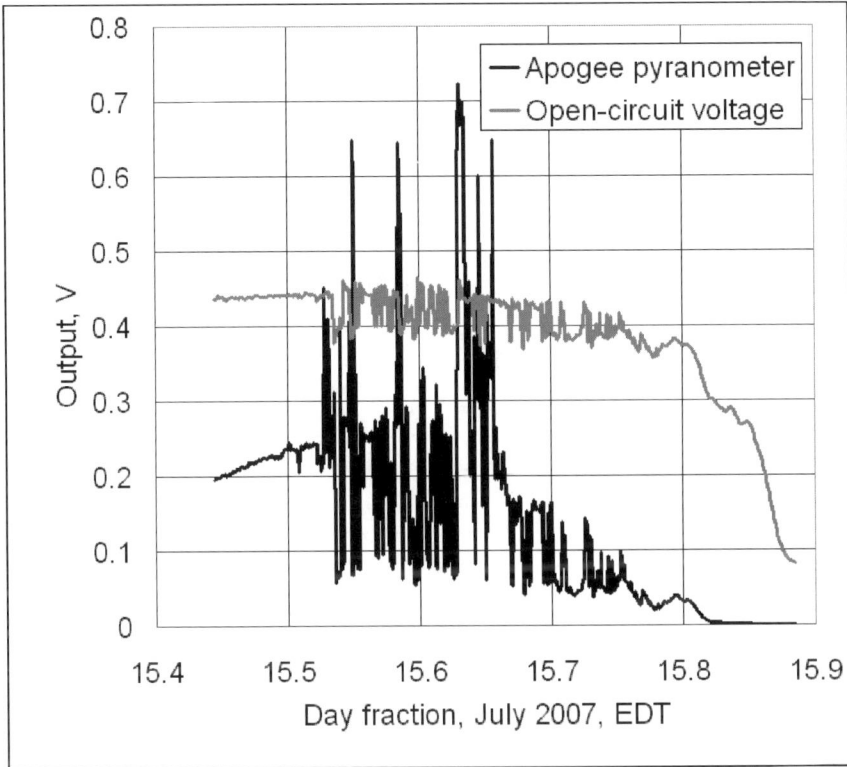

Figure 4.6. Open circuit voltage vs voltage proportional to current.

Figure 4.7 shows output voltage from the PDB-C139-based SIP described in this chapter, compared to output voltage from a commercial Apogee SIP. Ideally, these outputs should be linearly related, which allows a simple calibration of form

$$C\ [(W/m^2)/V]\cdot test\ (V) = reference\ [W/m^2] \qquad (4.2)$$

where the "reference" in this case is the Apogee output voltage multiplied by the manufacturer-supplied calibration constant.

With a 470-Ω load resistor, the relationship between outputs from the two instruments is very nearly linear, with a maximum output voltage of nearly 0.25 V in full summer sunlight. A quadratic curve fit to the data show that there is a little non-linearity, because the coefficient of the x^2 term is not exactly 0, but it is obviously much more linear than with a 940-Ω load.

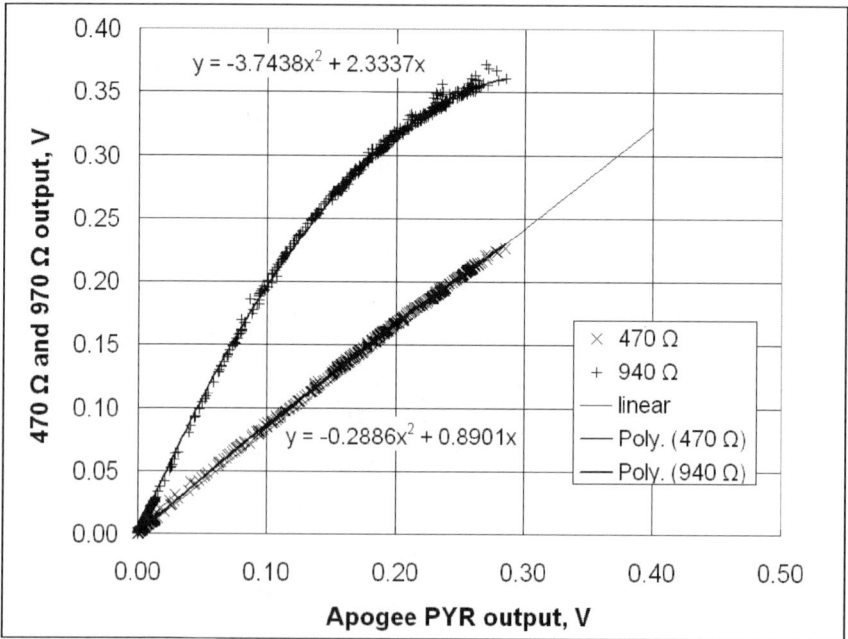

Figure 4.7. Output voltage of a PDB-C139-based pyranometer with two possible load resistances compared to output from a commercial pyranometer with a quadratic polynomial regression.

There is nothing "magic" about the choice 470 Ω as a load resistor—it is simply a metal film resistor that the author happened to have on hand when this instrument was being developed, and it turned out to provide a very good compromise between the competing demands of linearity and voltage output range. Metal film resistors offer a temperature coefficient (variation of resistance with temperature) that is about ten times smaller than for more common and less expensive carbon film resistors. They are therefore a much better choice for this application.

As expected, doubling the resistance increases the output voltage when insolation is relatively small but leads to a very nonlinear relationship between these two instruments when insolation is larger. Although in principle this nonlinear instrument response could be calibrated against the Apogee instrument, a linear relationship is a much better solution. It is reasonable to ask whether trying to apply a small non-linear calibration correction to the PDC-C139-based SIP with a 470-Ω resistor is worth the effort. For daily total insolation, for example, there is no justification for using anything more complicated than the linear calibration expressed in Equation (4.2). In practice, it is not possible to separate the small nonlinearity error from other, larger effects due to the pyrometer's imperfect cosine and spectral response.

Note, by the way, that there is no guarantee, and no way for the author or readers of this book to know, that the relationship between the Apogee pyrometer and the reference against which *it* is calibrated—a very expensive thermopile-based pyrometer—is completely linear with respect to insolation. For a relatively inexpensive commercial pyrometer whose absolute accuracy over a range of solar illumination conditions is no better than a few percent, this is a concern that perhaps can fairly be ignored.

The physical behavior of a photodetector forcing electrons through a load resistor, as shown in Figure 4.7, can be understood by using an analogy to highway traffic. The corresponding values are shown in Table 4.1. The lower the resistance value, the wider the highway. Current is analogous to the number of automobiles that pass along a highway per unit time and voltage is analogous to the number of cars per lane per second that pass along the highway.

Table 4.1. Electron flow and traffic flow.

Ohm's law → Traffic flow
I → cars/sec
V → cars/(lane·sec)
1/R → # of lanes
I = V/R →
(cars/sec) = [cars/(lane·sec)]·(# of lanes)

Ohm's law states that I = V/R. The traffic analogy—that the number of cars per second equals the number of cars per lane per second times the number of lanes—holds up as long as a highway can handle the traffic that is present. As more and more cars use the highway, if the flow of traffic is to be maintained, it is necessary to increase either the number of cars per lane per second, or the number of lanes. Physically, the number of cars per lane per second has a maximum value. Without more lanes, the flow of cars per second cannot be maintained—the highway becomes congested. This is analogous to the departure from linearity evident in the upper curve (970 Ω load) of Figure 4.7. Initially, the number of cars per lane per second is higher because the highway is narrower. If the highway is still sufficiently wide, traffic continues to flow freely with no reduction in "throughput" of cars per second, regardless of how many cars are using the highway. If there are more cars than the highway can handle, at some point the traffic will slow down and the number of cars per lane per second will no longer increase. In the case of Ohm's law, the properties of the resistor will change if too much power (current times voltage) is applied to it. It may melt or break—the equivalent of gridlock on a highway!

A completed PDB-C139-based SIP is shown in Figure 4.8. The detector with its Teflon® diffuser is at the right. The "bullseye" bubble level mounted on the left side of the 4 × 8-cm ABS plastic case is used to make sure the pyranometer is mounted level. The output cable terminates in a 2.5-mm stereo plug to fit the input jack on the logger used to record the data. (The middle ring on the stereo plug is not needed for the measurement, but it is required to be present, and un-connected, for use with the recommended data logger.)

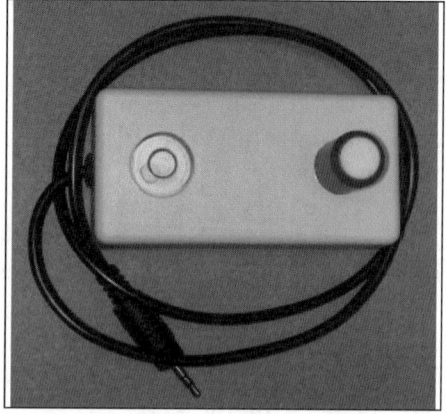

Figure 4.8. Completed PDB-C139 pyrano-meter.

The important message from this extensive discussion of solar cells and their use in pyranometers is that there are many decisions to be made in the design of even a very simple and inexpensive instrument, and the impact of all of these decisions must be investigated before an instrument can be relied upon to produce useful data.

4.1.2 Calibrating and Interpreting Pyranometers

Figure 4.9 shows the response of three different pyranometers: an Eppley Model 8-48 thermopile pyranometer, a LI-COR pyranometer (a widely used commercial SIP), and a very early version of a homemade SIP that used a thin-film amorphous silicon solar cell. This type of cell turned out to be a poor choice for a pyranometer. They are quite large (on the order of 10 cm^2), were not covered with a diffuser, and are prone to degradation under constant exposure to sunlight. Nonetheless, the data are still interesting because of the insight they provide into instrument calibration.

In Figure 4.9, the Eppley pyranometer (by far the most expensive) is taken as the reference. The LI-COR insolation values are based on the calibration supplied by the manufacturer. The homemade instrument is calibrated against the Eppley by minimizing the overall difference between the results over an extended period of time.

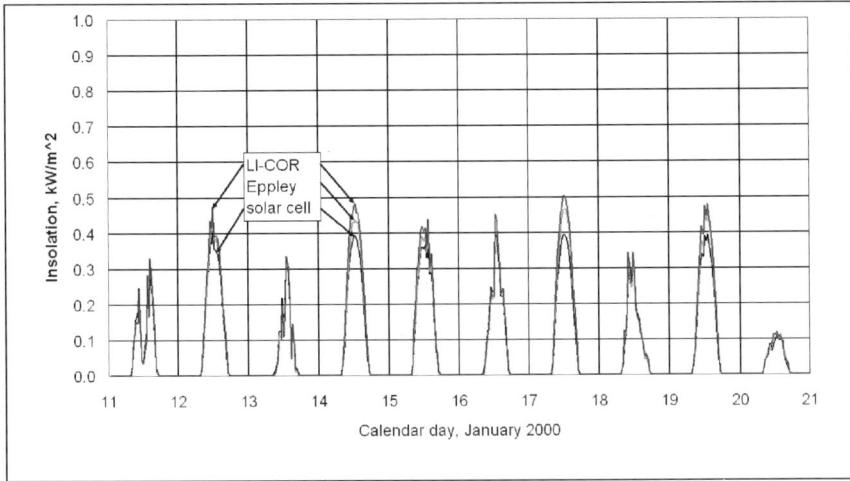

4.9a. Solar insolation during January 2000.

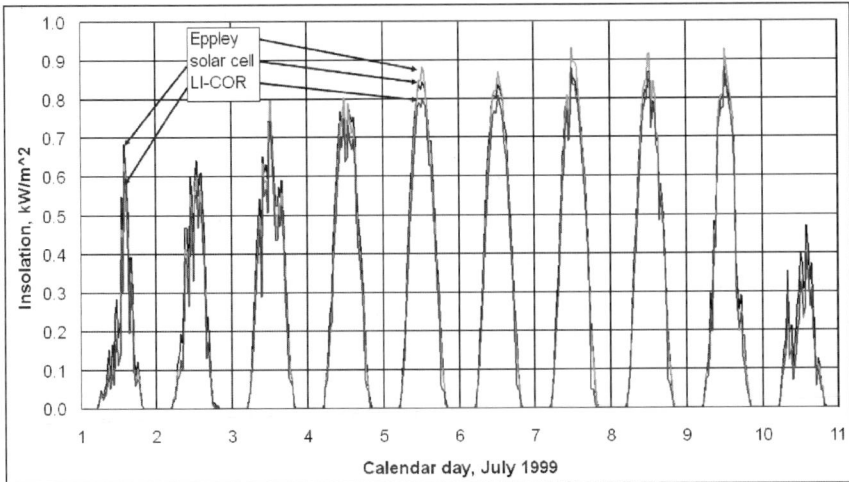

4.9b. Solar insolation during July 1999.

Figure 4.9 (see color plates). Solar insolation comparisons with three different pyranometers, Philadelphia, Pennsylvania, USA (lat = 39.96 N, lon = 75.19 W).

On clear days, the LI-COR values are higher than the Eppley values in the winter and lower in the summer. As currently calibrated, values from the homemade pyranometer are always a little less than those from the Eppley on clear days. The difference is larger during the winter than during the summer. There are several reasons why the calibrations cannot be made to match under all conditions. The cosine response of each instrument is different. Presumably, the Eppley's cosine response is better than either of the other two. The spectral response of the Eppley (a thermo-pile instrument) is different from that of the solar cell-based instruments. It is also likely that the response of the solar cell in this homemade pyrano-meter degraded somewhat between January and July, and that there is some nonlinearity in its response, as discussed in reference to Figure 4.7.

The peak response of the homemade instrument relative to the Eppley on clear days can, of course, be adjusted by changing the cali-bration constant. However, the differences in cosine and spectral response mean that better agreement on clear days will mean poorer agreement on cloudy days, because of cosine response differences and the fact that the sky is a different color on cloudy and clear days. The PDB-C139 SIP described in this chapter should perform better, certainly with less chance for performance degradation, than the one shown in Figure 4.9, but there are still inherent differences between thermopile-based and solar-cell-based pyranometers which will prevent these two kinds of instruments from agreeing under all sky conditions.

The differences between results from a high-quality thermopile pyranometer and a solar-cell-based "surrogate" pyranometer may already be small enough to yield sufficiently accurate data for some purposes, especially when insolation is averaged over a day. For other purposes, it may be possible to develop algorithms for variable calibration constants which depend on minimum solar zenith angle and the amount of "noise" in the insolation data. (Clear skies are less noisy than cloudy skies.)

An important point about pyranometers is that there is no definitive calibration source that is available to the amateur or educational user. The calibration of pyranometers against the highest quality research instruments is a major and ongoing project at the National Renewable Energy Laboratory's Solar Radiation Research Branch in Golden, Colorado, USA. Two samples of the PDB-C139-based pyranometer des-cribed in this chapter were included in the 2007 Broadband Outdoor Radiometer Calibration (BORCAL) project. A photo of the calibration setup at NREL is shown in Figure 4.10. Other instruments included an Apogee PYR-P, an SP Lite silicon solar-cell instrument from Kipp & Zonen, and a high-end CM-22 thermopile pyranometer from Kipp & Zonen. At the time this book was written, the PYR-P cost about $170, the SP Lite about twice as much as the PYR-P, and the CM-22 about $6,500.

The BORCAL calibration procedure provides an average responsivity for each calibrated instrument, in units of $\mu V/(W/m^2)$ (the inverse of a calibration constant as defined above in Equation (4.2)), but also a set of solar zenith angle dependent responsivities. These are not direct measures of cosine response, as the responsivities are determined

Figure 4.10. PDB-C139 SIPs being calibrated at NREL.

under full-sky conditions by comparing output to reference thermopile pyranometers with very good cosine response, but they do account for the effects of cosine response deficiencies under clear sky conditions. Such data, which would otherwise be very difficult to obtain, can be used to improve the accuracy of pyranometer measurements under some sky conditions.

Figure 4.11 shows the zenith angle dependent responsivities for a PDB-C139 SIP and an Apogee PYR-P. Clearly, the Apogee instrument, which costs roughly ten times as much as the PDB-C139 SIP instrument, has a better cosine response. Its responsivity is very nearly linear out to a solar zenith angle of about 75° but increases rapidly after that. The PDB-C139 responsivity is less linear and starts to decline after about 55°. This zenith angle dependence looks more serious than it actually is for most purposes, because of the decrease in insolation as a function of zenith angle. Both the PDB-C139 and PYR-P instruments exhibit a little azimuthal asymmetry, with responsivity differences between morning and afternoon. At least for the PDB-C139, this asymmetry is explained by the fact that the active detector area is square rather than circular, and not mounted precisely in the center of its housing. (Refer to Figure 4.4.) The lower limit on solar zenith angle in the responsivity data (about 18°) is set by the latitude of the BORCAL site (39.7° N); these missing data are inconsequential for these instruments.

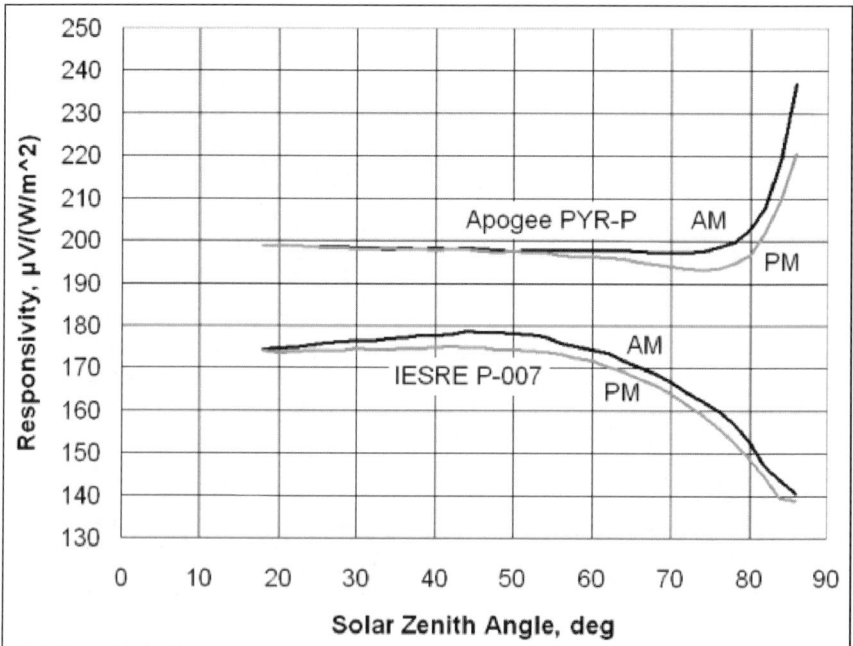

Figure 4.11. Solar zenith angle responsivities of PDB-C139 (IESRE P-007) and Apogee PYR-P pyranometers, based on 2007 BORCAL results.

The impact of zenith angle dependent responsivity on retrieved insolation values depends in part on how the data will be used. In the field, imperfect cosine response is affects output along with other effects, including cloud cover (which alters cosine response effects), variable water vapor, and poor air quality (which reduces "clear sky" insolation by an amount that requires other kinds of measurements to quantify). The cumulative impact of these effects can easily overwhelm cosine response effects. Because the spectral response of SIP pyranometers is so different from thermopile pyranometers, it is not possible to eliminate all discrepancies between these two basic pyranometer designs.

Another alternative for calibrating pyranometers is to use models of insolation that can be applied on cloud-free days when the atmosphere is clean and dry. A simple model is given in Appendix 5, which also includes references to other models. Of course, a procedure that requires knowledge of the insolation to calibrate an instrument whose purpose is to measure insolation sounds more than a little suspect! However, the fact remains that it is possible to calculate insolation quite accurately under ideal sky conditions. Once this is done, then the instrument can be used to measure insolation under less than ideal conditions.

There is one final performance/calibration consideration for pyranometers: all photodetectors exhibit some temperature sensitivity, and the detector used here is no exception. In the online literature about its PYR-P pyranometer, Apogee Instruments states that "the temperature response is less than 0.1% per degree Celsius" [Apogee Instruments, 2008]. The extent to which this estimate is based on actual measurements in the field, rather than on published information about the photodetector used in their instrument, is unknown. Calibrations are typically done at temperatures around 20–25°C. The output of a silicon photodetector decreases as the temperature increases, so the calibration constant, as defined in 4.2, must be increased for detector temperatures above 20–25°C and decreased for temperatures below 20–25°C.

If you know the temperature sensitivity of a SIP, a temperature correction is easy to apply except for this detail: the temperature you need to know is not the ambient air temperature, which is easy to obtain, but the actual temperature of the detector, which is not at all easy to obtain!

For the PDB-C139 SIP, detector temperatures can be approximated by measuring temperature inside an identical housing. Such a device is shown in Figure 4.12. The LM35DZ integrated circuit produces an output voltage directly proportional to Celsius temperature—10 mV/°C—so a temperature of 30°C produces an output of 0.30 V. This simple circuit is very easy to build and use and responds quickly to temperature changes. (You can try it by blowing on the detector in a cool room.) It draws very little current (about 60 µA), and it doesn't need a regulated supply voltage.

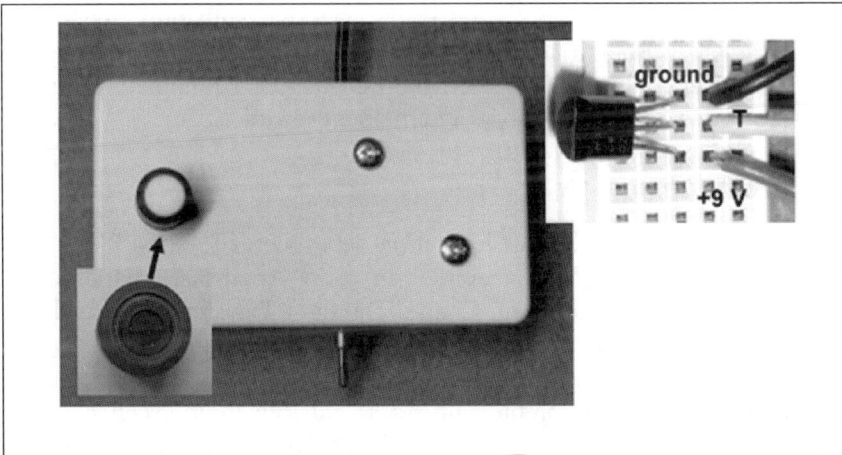

Figure 4.12. A simple instrument for estimating pyranometer detector temperature sensitivity.

It can be powered with a single 9-V battery and its output can be recorded along with pyranometer outputs. The logger requires non-negative voltage inputs, so the temperatures must be above freezing.

Some preliminary measurements with this device demonstrate that it works as planned. However, determining a reliable temperature correction requires knowing the actual insolation based on data from a high-quality thermopile pyranometer, and this is an expensive proposition that could not be undertaken during the preparation of this manuscript.

In any event, ongoing comparisons of the very inexpensive pyranometer described in this chapter against data from Apogee pyranometers reveal no differences that can be systematically related to temperature. This can be interpreted to mean that the temperature sensitivity of these two instruments is roughly the same. Whatever temperature dependence remains cannot be removed without additional equipment and measurements under a wide range of temperature conditions.

Questions that remain about these inexpensive pyranometers concern their long-term stability and reliability. Pyranometers and other solar monitoring instruments that are permanently mounted outdoors face a very harsh environment, including continuous temperature cycling and UV exposure, which can seriously degrade materials. This is a potential problem even for very expensive instruments. Testing side by side with Apogee pyranometers during 2007 demonstrated that the PDB-C139 SIPs performed very well. In two cases, the voltage output of Apogee pyranometers decreased by nearly 30% in the space of a few days, for reasons that were never determined, while the PDB-C139 instruments continued to perform as expected. As a result, at one site, with only an Apogee pyranometer, a usable data record was lost. At the other site, with both instruments operating side by side, the PDB-C139 pyranometer provided the redundancy needed to ensure a permanent record.

4.1.3 Applications

Figure 4.13a shows 8 days of insolation data recorded with a PDB-C139 SIP during May 2007, at a school in rural Arkansas. The diamond symbols placed near 1:00 p.m. Central Daylight Time indicate insolation values calculated with a clear sky model (see Appendix 5), using typical atmospheric conditions. The insolation values recorded around solar noon, a little over 1,000 W/m^2, on the 13th, 14th, and 16th are in excellent agreement with this model. Other days are cloudy, and under such conditions, reflections from the sides of clouds can produce insolation values temporarily larger than those that would be observed on a clear day. The low insolation readings on the afternoon of the 15th, approaching 0, indicate

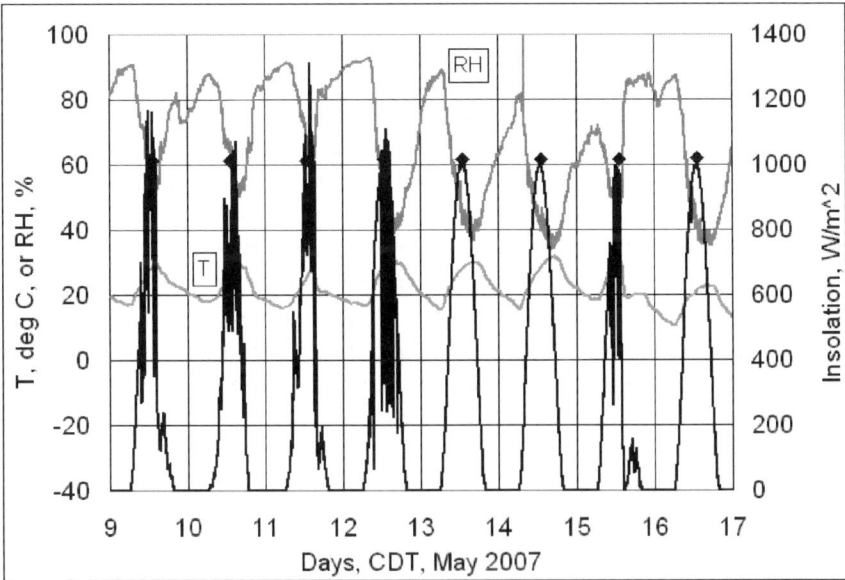

4.13a. Insolation, temperature, and relative humidity at 1-min intervals.

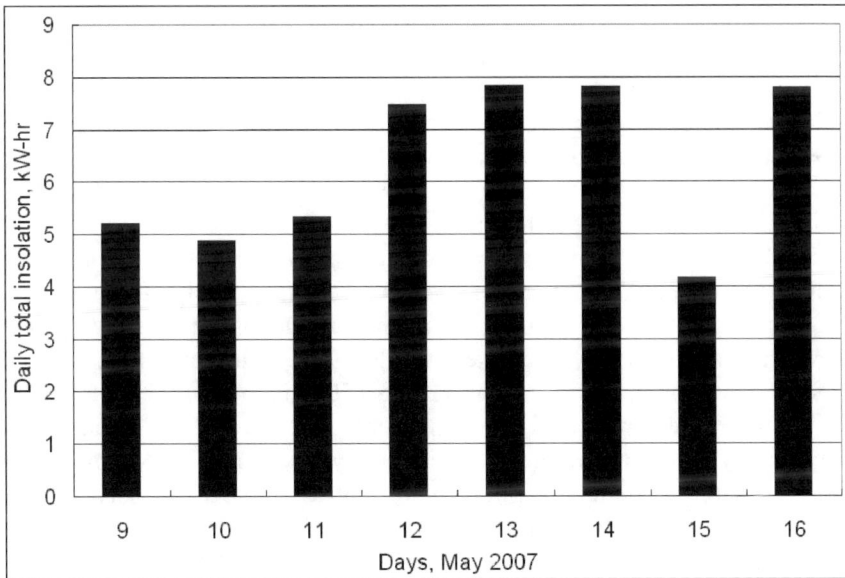

4.13b. Daily insolation, kW-h.

Figure 4.13. Insolation data from Arrie Goforth Elementary School, Norfork, Arkansas, USA, 9–16 May 2007. Latitude 36.1972° N, longitude 92.2688° W [Data provided by Wade Geery].

very heavy rain under dark, overcast skies. These data show clearly the daily cycles in temperature and relative humidity, and the inverse relationship between these two quantities.

The unit of measurement in Figure 4.13a is W/m^2. In Figure 4.13b, the unit is kW-h per day. This value is calculated by summing the insolation values over the day and dividing by 60 min per hour and by 1,000 to convert from watts to kilowatts. Note that despite the momentarily high insolation values due to reflection from clouds, the most power integrated over a day is still delivered on the clearest days.

Regardless of how expensive a pyranometer is, its performance can change over time in the harsh outdoors environment. The output of detectors can degrade. The transmission of sunlight through a diffuser can diminish due to color changes caused especially by UV radiation. The Teflon® used in the PDB-C139 diffusers is actually somewhat porous, and some researchers believe that very small dirt particles can become permanently embedded in this material.

Pyranometer manufacturers always recommend that their instruments be recalibrated regularly (for example, every 2 years), a recommendation that is often not taken as seriously as it should be because it can be expensive and disruptive to an ongoing measurement program. For a serious program of insolation measurements, the solution is to have multiple pyranometers at the same observing site, including at least one reference instrument that is used only sporadically for calibration checks, rather than being continuously exposed to sunlight and weather. The inexpensive pyranometers described here actually make a proactive maintenance and calibration program more likely, because the low per-instrument cost encourages redundancy.

It might be tempting to conclude that the effort described in this section to understand the performance of a very inexpensive instrument is excessive and that it is unfair, or even a waste of time, to judge the performance of a $10 pyranometer (roughly the cost of parts) against even a $170 instrument, not to mention a $6,500 instrument. However, an essential requirement for all the instruments described in this book is that they must yield reliable and usable data. Any instrument that cannot meet this standard is better forgotten, no matter how inexpensive it is. Fortunately, the BORCAL project has presented a unique opportunity to document the performance of the PDB-C139 SIP described here relative to instruments that are widely used and accepted within the science community. This kind of "pedigree" for such an inexpensive instrument is unusual, but obtaining it has certainly been worth the effort!

4.2 Using Light-Emitting Diodes as Inexpensive Spectrally Selective Detectors

Each of the instruments discussed in this book shares a common underlying physical principle. When solar radiation strikes a light-sensitive detector (a photodetector), atoms in the detector absorb some of the energy from photons. In this excited state, which may be produced by light over a broad range of wavelengths or over only a very specific range of wavelengths, the atoms release electrons, which can flow through a conductor to produce an electrical current. This current is assumed to be proportional to the intensity of the radiation striking the detector.

As shown in the previous section, this concept is easily implemented in an inexpensive pyranometer, which measures broadband solar radiation. However, measurements that must be made over just part of the solar spectrum pose additional problems. Recall, for example, the discussion of photosynthetically active radiation (PAR) in Chapter 3. PAR instruments should measure only visible light between 400 and 700 nm. The PDB-C139 pyranometer detector is clearly unsuited as a PAR detector because (recalling Figure 4.2) its response is strongly peaked in the near-IR.

As discussed in Chapter 3, commercial spectrally selective instruments typically use various kinds of light filters to restrict the response of broadband detectors. The fact that these filters are expensive, fragile, and subject to unpredictable degradation has led to the development of instruments that use LED detectors. This application was first described in the peer-reviewed literature by Mims [1992].

It is easy to demonstrate that LEDs respond to light simply by making the same measurement on an LED that was previously made on a solar cell. (Recall Figure 4.3.) Table 4.2 shows the open-circuit voltage and short-circuit current for several LEDs exposed to sunlight.

Table 4.2. Response of various LEDs to sunlight.

Part ID	Color	Open-circuit voltage (V)	Short-circuit current (μA)
HLMP 3762	Red	0.138	0.1
HLMP D600	Emerald green	0.780	0.8
HLMP C30	Blue	0.077	0.1
F5E1	Near-IR	0.903	9.2

Note that although the open-circuit voltage is reasonably large, the short-circuit current is very small—on the order of a μA instead of the several tens of mA produced by a small solar cell. Even the 2 × 2 mm PDB-C139 pyranometer detector produces ~0.5 mA in full sunlight, about two or three orders of magnitude more than these LEDs. This discrepancy between solar cells and LEDs should not be surprising because a solar cell is optimized specifically to generate electrons when light strikes it, whereas an LED is optimized to work the other way around—to generate light when electrons are passed through it.

The values in Table 4.2 are representative, but they would be different for other LEDs measured under the same circumstances. (These measurements were taken on a very hazy late summer day.) There is no particular pattern relating emission color to either open-circuit voltage or short-circuit current. These values are a function of the chemistry and physics of the LED chip itself, and there is no reason to suspect that the values in Table 4.2 could be used to predict the behavior of other LEDs, for example. Also, the current values, especially the 0.1 μA values, are not very accurate, as the measurements were made with a digital multimeter set to its 200 μA scale, which reads between 0.1 and 199.9 μA. The purpose of Table 4.2 is simply to make the point that, although the performance of solar cells and LEDs when exposed to sunlight is very different, the operating principles are identical.

Because the current output of LEDs is so small, it is not possible to measure and record their current output directly and accurately with inexpensive equipment. If you were to "load" an LED with a resistor, as was done previously for recording the output of a solar cell pyranometer, you would not obtain a voltage level suitable for recording with an inexpensive data logger. This makes sense because these LEDs are very small devices that are not intended to do enough "work" to drive such a logger. This problem is easily solved by amplifying the current and converting it to a voltage with a simple electronic circuit called a transimpedance amplifier. Expensive data loggers intended for use with photodetectors are no different in principle from inexpensive ones, but they may include a built-in transimpedance amplifier just for this purpose. Transimpedance amplifiers suitable for using with LEDs are inexpensive and easy to build using an operational amplifier (op amp) with only a few additional components. Appendix 7 gives more details about building transimpedance amplifiers.

It would be very easy to select LEDs for use as light detectors except for the fact that the spectral distribution of an LED's response to light is not the same as its emission spectrum. The latter is readily available from the manufacturer, because this is the fundamental information required to select an LED for use as a light source in a particular application. Information about the response spectral distribution is rarely available, because it is typically of no interest to the purchaser of an LED.

So, the response spectrum must be measured in order to determine the suitability of an LED as a light detector for a particular application.

The peak response wavelength is invariably lower than the peak emission wavelength, but as a practical matter, it is not possible to predict the peak wavelength and shape of an LED's response spectrum based on its emission spectrum. Figure 4.14 shows the normalized generic emission spectrum of a typical emerald green LED, as supplied by its manufacturer, compared to its normalized directly measured response spectrum.[1] According to these data, the emission peaks at about 555 nm, but the peak response is around 530 nm.

It is not difficult to measure the response spectrum of an LED in the near-UV, visible, or near-IR. However, the required equipment costs a few thousand dollars, which places these measurements beyond the reach

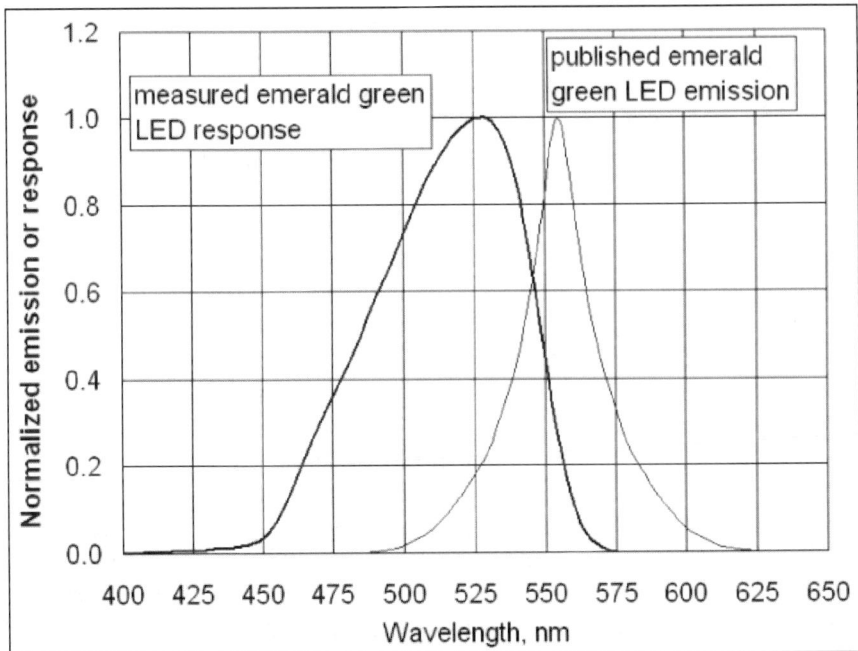

Figure 4.14. Emerald green LED emission compared to measured spectral response.

---

[1] An "emerald green" LED is used because its spectral response bandwidth is typically narrower than that of other "green" LEDs. This is desirable for sun photometers, as discussed in Chapter 5.

of most readers of this book. Consequently, some specific information about detectors will need to be provided for instruments that require responses at particular wavelengths.

4.3 Measuring Photosynthetically Active Radiation

4.3.1 Designing an LED-Based PAR Detector

There is no need to design expensive thermopile-based detectors for photosynthetically active radiation because of the limited spectral response that is required—the visible part of the spectrum from 400 to 700 nm. Commercial PAR detectors use filtered silicon-based detectors for this task even though, as previously shown, the response of such detectors is not flat even across the visible spectrum.[2]

It has already been noted that the pyranometer detector described in the previous section is not a good PAR detector because its spectral response must be filtered. An alternative is to use one or more LED detectors with responses in the visible spectrum. Figure 4.15 shows a typical transmission spectrum for a green leaf. Leaves reflect and transmit solar radiation in the green part of the visible spectrum, and use radiation in the rest of the visible spectrum for photosynthesis. Therefore, it would be useful to use LEDs that detect sunlight at wavelengths on either side of green light—one that detects red light and another that detects blue light, as shown in Figure 4.15 [Mims, 2003a].

The silicon carbide (SiC) LED whose response is shown in Figure 4.15 is no longer available and even if it were, its peak response wavelength lies in the ultraviolet spectrum below 400 nm.[3] So, a two-LED PAR detector may not be practical. However, even a single LED that responds to light at the red end of the visible spectrum produces an output that correlates well with commercial PAR detectors. Such LEDs are widely available and very inexpensive, much less than $1. For this application, which does not depend on the theoretical monochromatic light assumption that applies to direct sunlight instruments, the wider spectral response of

---

[2] These filters, which ideally have a square spectral response—flat between 400 and 700 nm and zero everywhere else—are called "top hat" filters.

[3] It is reasonable to hope that "aqua" LEDs, which emit bluish-green light rather than blue light, would have a somewhat higher response wavelength, but this is not true. They also respond to ultraviolet light rather than visible light.

gallium phosphide (GaP) LEDs is actually a better choice than the narrow response of aluminum gallium arsenide (AlGaAs) LEDs.

A 4-year series of data from a commercial PAR sensor from LI-COR Biosciences and a PAR sensor using the two detectors shown in Figure 4.15 is shown in Figure 4.16 [Mims, 2003]. Long-term comparisons with a single red LED detector are not yet available.

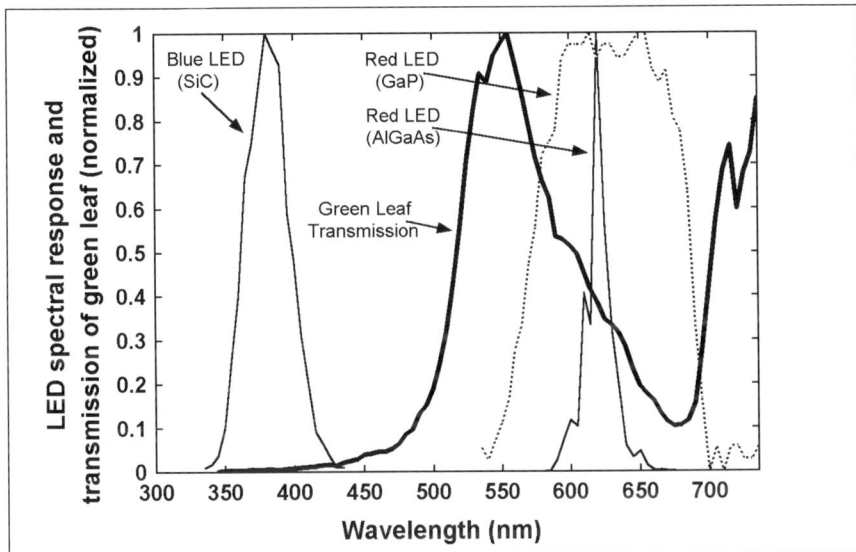

Figure 4.15. Normalized leaf transmission and LED responses for monitoring photosynthetically active radiation [Mims, 2003].

4.3.2 Calibrating and Interpreting a PAR Detector

The interpretation of PAR measurements may be less intuitive than insolation measurements because the units are photons per unit area (see Section 3.3.3) rather than watts per unit area. Insolation measurements are typically taken in an open area with an unobstructed view of the horizon. PAR measurements can also be taken in an open area, but it is also interesting to measure PAR under a variety of other conditions, including under vegetation. The "color" of incident radiation is different under vegetation than under open sky, so in these circumstances, differences among detector spectral responses becomes very important.

There is no readily available calibration method for an LED-based PAR detector other than comparing its output to a commercial instrument, as has been done in Figure 4.16. Because the definition of

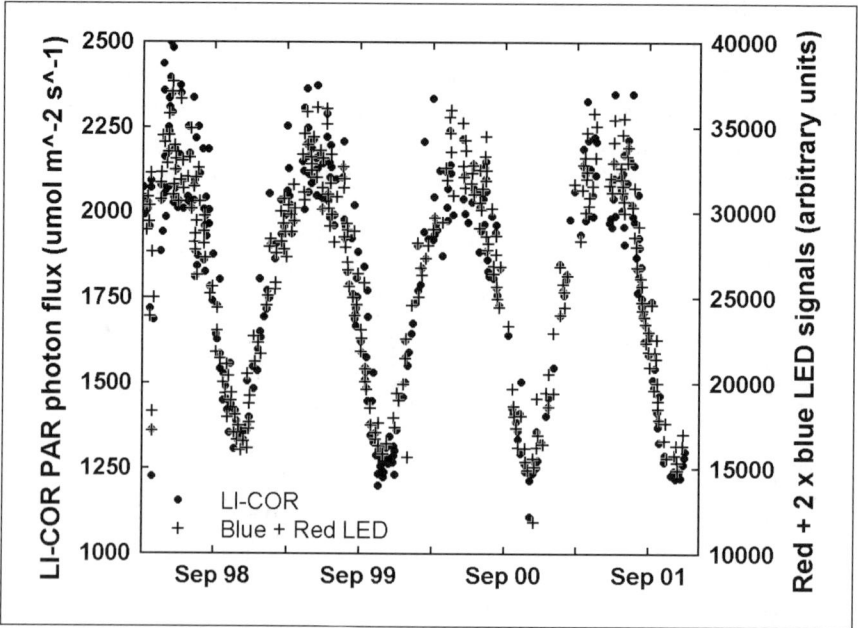

Figure 4.16. Time series of solar noon observations of PAR at Seguin, Texas, USA, measured by a commercial PAR sensor and by combining outputs from a blue-red LED pair [Mims, 2003].

photosynthetically active radiation is somewhat arbitrary, PAR measurements are essentially defined by a particular instrument. If practitioners accept LI-COR PAR radiometers for PAR measurements, as appears to be the case, then that instrument becomes the accepted reference. In a sense, that makes the design of inexpensive instruments easier. If they can be made to agree reasonably well with a LI-COR PAR radiometer, then they are, by definition, calibrated.

4.4 Measuring Ultraviolet Radiation

4.4.1 Designing a UV Radiometer

It is not easy to measure UV radiation accurately, because the total energy in the UV part of the spectrum is small and because suitable detectors tend to be expensive. However, blue-emitting LEDs, which detect light in the UV, provide opportunities for making some kinds of UV measurements with relatively inexpensive instruments.

The normalized spectral response of an HLMP-C30 blue LED is shown in Figure 4.17. It has a strong peak response around 372 nm, in the UV-A part of the spectrum. Unfortunately, it also has a significant response "shoulder" that extends into the blue part of the visible spectrum beyond 400 nm.

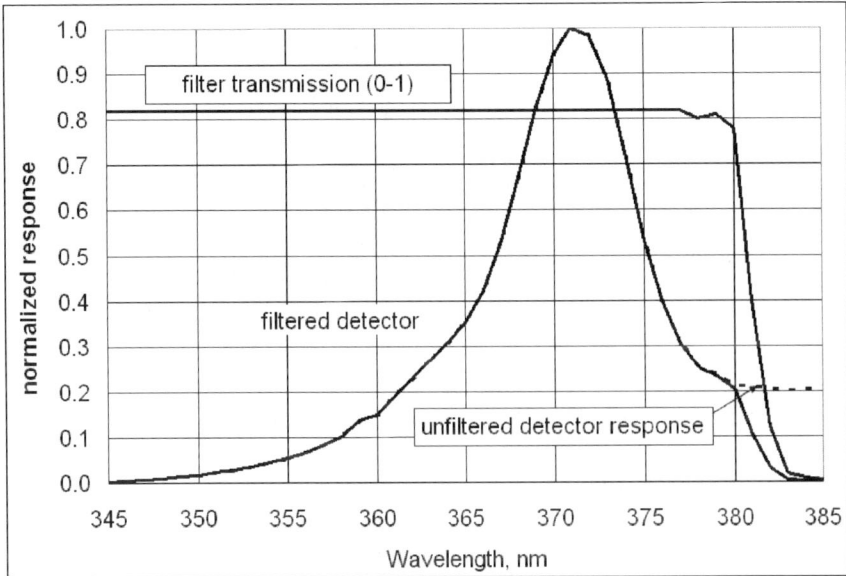

Figure 4.17. Spectral response of a filtered HLMP-C30 blue LED.

The solution to this problem is to cover the LED with a "low-pass" filter—a coated piece of glass or quartz that is highly transparent to wavelengths below about 380 nm, but does not transmit any radiation at longer wavelengths, over the visible part of the spectrum.

A cut-away view of a machined nylon housing with the blue LED and its filter, which is 7 mm in diameter, is shown in Figure 4.18. These kinds of filters must be custom-made for a specific application and are therefore quite

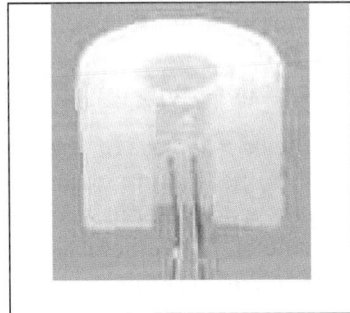

Figure 4.18. Cut-away view of a UV-A detector assembly, without its Teflon® diffuser.

expensive. They can be manufactured economically only in large batches, on the order of a minimum of 100 units, with a per-unit cost roughly ten times the $1–2 cost of the LED detector itself. These filters can degrade, and it is necessary to monitor their performance carefully. As a result, the UV-A radiometer is by far the most expensive and challenging instrument discussed in this book.

The LED detector assembly is covered with a Teflon® disk to diffuse sunlight, similar that used for the pyranometer. This assembly is then attached to the top of the instrument enclosure. A completed UV-A radiometer is shown in Figure 4.19. The dark ring around the detector housing, machined from a piece of thick-walled "Schedule 80" PVC plumbing pipe, prevents light from leaking through the sides of the housing. The ring can be removed and replaced with a col-limating tube with a small hole at

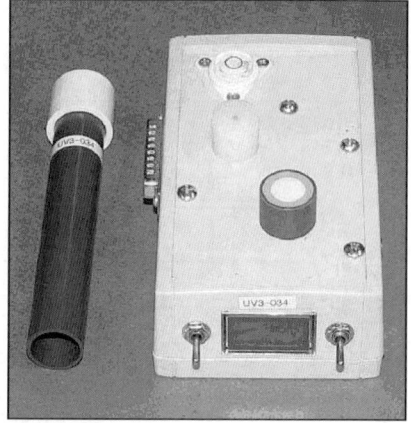

Figure 4.19. UV-A radiometer with optional collimating tube.

one end (the end with the white cap in Figure 4.19). This tube, which is also made from PVC plumbing pipe and a standard end cap fitting, fits snugly over the housing and gives the detector a field of view identical to the sun photometers that will be discussed in Chapter 5. Thus, a single instrument can be used to measure either full-sky or direct solar radiation.

There is a bubble level mounted at the rear of the case. The other housing visible on the top of the case contains an LM35DZ temperature sensor whose output is assumed to represent the temperature of the LED detector in its housing.

4.4.2 Calibrating and Interpreting Data from a UV-A Radiometer

The peak response of this UV-A radiometer, at about 372 nm, is very close to the 368-nm channel on the Yankee Environmental Systems UV Shadowband Radiometer, a widely used multi-channel research-quality radiometer. Thus, the instrument described here can be calibrated, in units of W/m^2, against this much more expensive instrument.

The first potential concern for this instrument is its cosine response, as discussed for pyranometers earlier in this chapter. There is no reason to expect the cosine response of these inexpensive instruments to be as accurate as that of more expensive instruments. However, cosine response corrections for the UV-A radiometer described turn out not to be very important for many purposes.

Figures 4.20a and 4.20b show some data for two UV-A radio-meters located at NASA/Goddard Space Flight Center and a reference radiometer at a nearby U.S. Department of Agriculture's research site in Beltsville, Maryland. Both are part of the USDA UV Monitoring Network. Figure 4.20a shows calibrations of instruments #027 and #030 against the reference instrument with no cosine correction. Figure 4.20b shows the same data calibrated with a cosine correction applied to the test instruments.

Considerable effort is required to measure the cosine response of radiometers under realistic conditions. UV radiometers pose additional problems because of the expense of equipment that will produce enough UV radiation to roughly approximate sunlight. Even under clear sky con-ditions, some theoretical calculations are required to model the contributions of direct and diffuse radiation to the total output of a detector. So, it is fortunate that even the simplest calibration strategy, which merely assigns a single coefficient by which instrument voltage outputs are multiplied to convert the output to units of W/m^2, works reasonably well. In the un-corrected data of Figure 4.20a, there are only some small residual visible differences relative to the reference instrument in the afternoon. (These differences are hardly visible in Figure 4.20.)

The second potential concern for such instruments is their temperature sensitivity. Any calibration strategy must assume either that this dependence is too small to significantly impact a specific application, or that it can be appropriately characterized and accounted for. Expensive radiometers use heaters to actively maintain the detector at a constant temperature, typically 40°C, but this is not a practical solution for an inexpensive battery-powered instrument.

Figure 4.21 shows uncalibrated voltage outputs from an early prototype UV-A radiometer, a first attempt to look at possible temperature sensitivity. The data were collected in late morning, local time, under a partly cloudy and therefore very "noisy" sky, as indicated by the UV-A detector output in the bottom trace. (The time axis is Universal Time in minutes (1,440 min per day), 4 h later than Eastern Daylight Saving Time.) The upper trace is output from an LM35DZ temperature sensor whose output is equal to the temperature in degrees Celsius divided by 100. On this occasion, the sensor temperature rose by about 15°C over 20 min. However, there is no correspondingly large rise in output; the slow rise in the UV-A output level over the entire data collection period is due to the fact that the sun is still rising in the sky. Still, temperature sensitivity is a

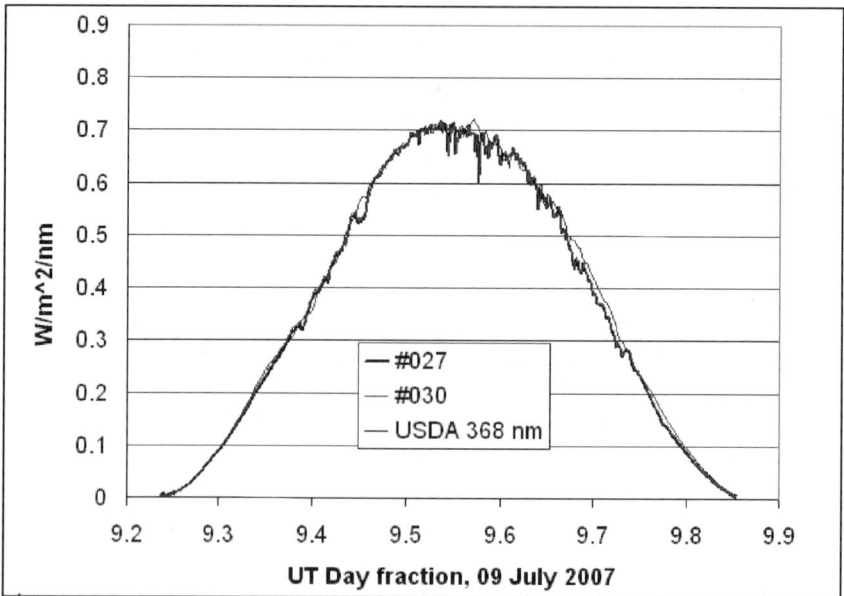

4.20a. No cosine correction.

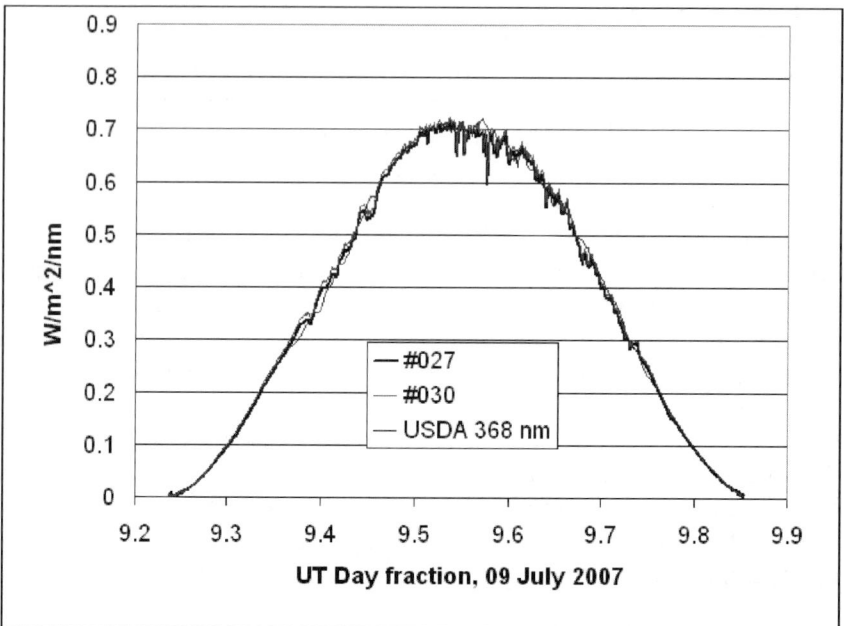

4.20b. With cosine correction.

Figure 4.20 (see color plates). Two UV-A radiometers calibrated against a USDA
UV site in Beltsville, Maryland, without and with cosine correction.

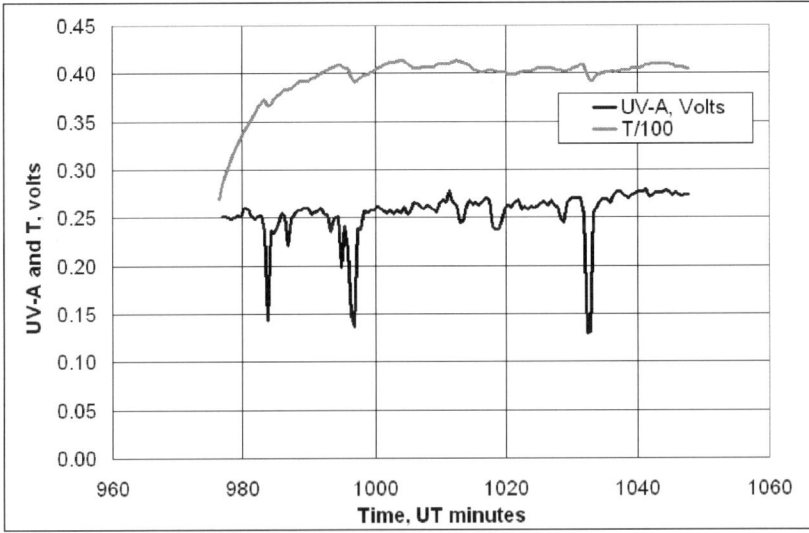

Figure 4.21. Detector and temperature outputs from a UV-A radiometer, 13 May, 2004.

matter that needs to be examined over a wide range of conditions whenever characterizing the performance of any radiometer.

Table 4.3 shows some calibration data for the same two UV-A radiometers whose calibration against a reference radiometer was shown above in Figure 4.20. The temperatures are maximum daytime temperatures—not air temperatures, but the average of the two maximum temperatures recorded by the LM35DZ sensors mounted on each radiometer case. The assumption is that this temperature is close to the actual detector temperature.

The plot of the data in Table 4.3, shown in Figure 4.22, exhibits considerable scatter. The most obvious explanation is that the strategy for determining the actual detector temperature is not very accurate. However,

Table 4.3. Calibration data for two UV-A radiometers.

Date	T (°C)	Calibration constants		Reference
		#027	#030	
26 Sept. 06	35	1.45	1.28	MD12 388
17 Jan. 07	8	1.74	1.52	MD12 397
08 Feb. 07	10	1.74	1.52	MD12 397
23 Apr. 07	39	1.46	1.34	MD12 397
02 July 07	38	1.55	1.40	MD12 388
09 July 07	48	1.45	1.32	MD12 388

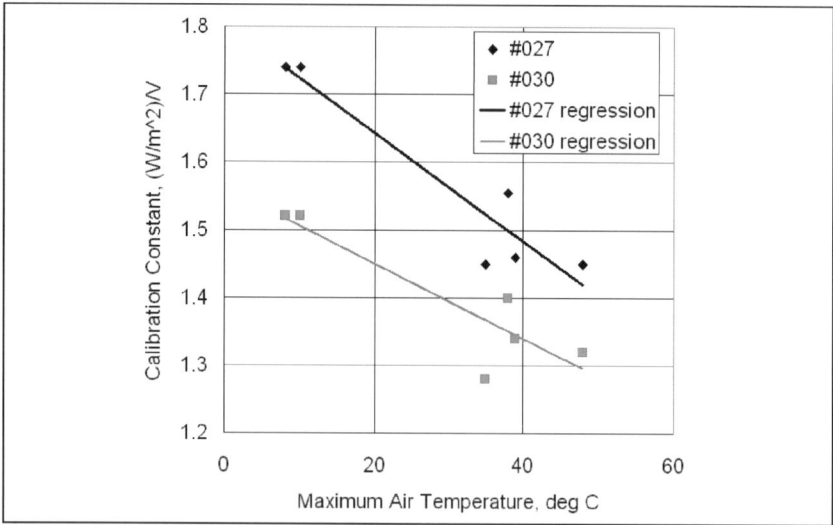

Figure 4.22. Temperature sensitivity of the calibration constant for two UV-A radiometers, compared to a reference instrument.

these temperatures are also a function of the time of year. The minimum zenith angle is, of course, also a function of the time of year, which means that temperature effects on calibration are intertwined with cosine effects. It is not easy to decouple these two effects under realistic conditions in the field! Nonetheless, these experiments demonstrate that it is both necessary and possible to provide temperature corrections to the output from these instruments.

It might seem reasonable to conclude that the uncertainties in the calibration of these inexpensive instruments render them unsuitable for doing "real" science. However, this is not true! As noted at the beginning of this section, accurate UV measurements are *very* difficult to make. Even the most expensive, independently calibrated UV radiometers can produce results that vary by a few percent when they are compared side by side. Constant monitoring of instrument performance and ongoing recalibrations are a way of life for scientists who need these kinds of data. Fortunately, a great deal of the science interest in UV radiation concerns its spatial and temporal variability. With rigorous ongoing programs of instrument intercomparisons, it is possible to address many interesting research questions even when uncertainty remains about absolute radiometric accuracy.

The original motivation for developing this UV-A radiometer was to support some of the data products resulting from measurements made by instruments on NASA's EOS/Aura spacecraft. These measurements

include UV radiation at Earth's surface. The effects of clouds and stratospheric ozone on UV radiation reaching the surface seriously impact the
calculations required to develop these data products, and ongoing ground-
based measurements are required in order to validate the performance of
data-processing algorithms for space-based measurements. Direct sunlight
and the optical depth at UV-A wavelengths are important because there is
no significant ozone absorption at these wavelengths. The instrument
described in this section allows these ground-validation measurements to
be made at a much higher spatial density within the field of view of Aura
instruments than would be feasible with much more expensive instruments.

4.5 A Surface Reflectometer

4.5.1 Designing a Two-Channel Reflectometer

Any of the full-sky instruments described in this chapter can be used to
measure surface reflectance simply by using two instruments—one pointing up and the other pointing down. Figure 4.23 shows one interesting and

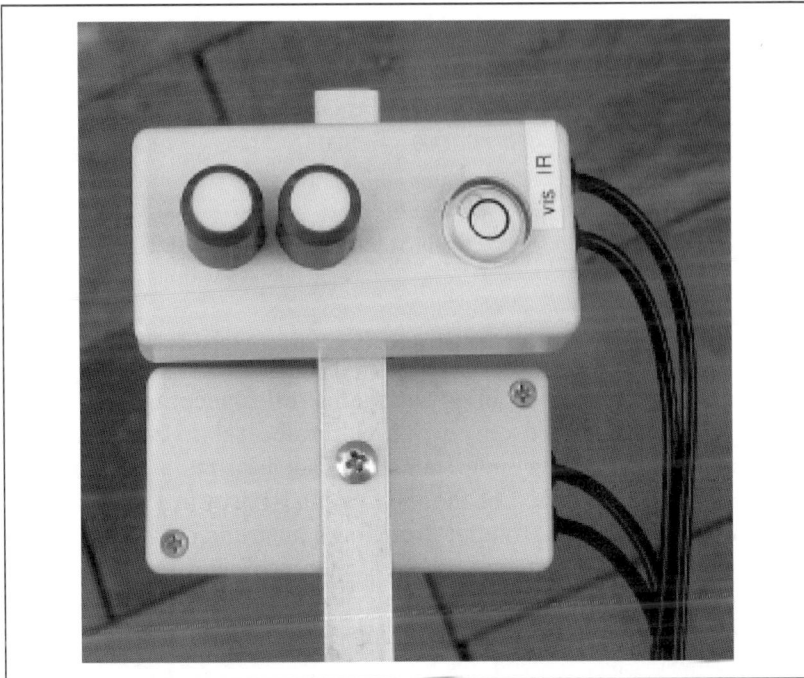

Figure 4.23. A reflectometer that uses two identical two-channel radiometers.

inexpensive configuration. The PDC-C139 detector is paired with a companion detector that measures only infrared light (PDB-C139F). These two detectors are identical except that the "F" version is in a black T 1-3/4 housing that blocks all visible light but is transparent to radiation in the near-IR. Two of these two-channel instruments are mounted at one end of a 2-m piece of square aluminum tubing.

The relative spectral responses of these broadband and near-IR detectors are shown in Figure 4.24, based on data provided in the manufacturer's data sheets.

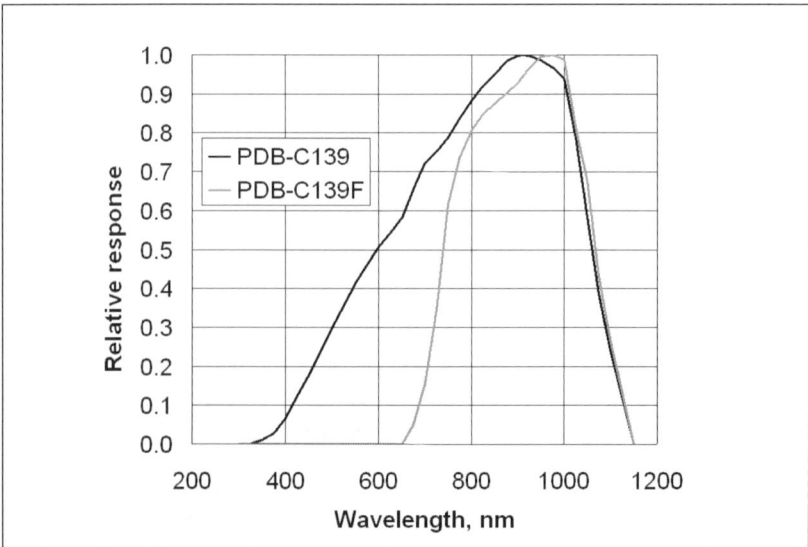

Figure 4.24. Normalized response for broadband and near-IR silicon-based photo-detectors.

4.5.2 Calibrating and Interpreting Reflectance Measurements

As described in Chapter 3, an advantage of reflectance measurements is that the instruments need only to be calibrated relative to each other, rather than against an absolute standard. That is a major advantage for most readers of this book. For the instruments shown in Figure 4.23 it is sufficient just to log the output from both instruments when they are pointing up. One is designated as the "reference" and the other is calibrated to that reference. It might also be a good idea to repeat this relative calibration with both instruments pointing down.

Figure 4.25 shows some reflectivity measurements taken with the device shown in Figure 4.23. These two instruments were calibrated relatively to each other as described in the previous paragraph. The data were collected by walking with the reflectometer over three different surfaces—flagstones, gravel, and grass, and back again. Note that only the output voltages are needed, rather than physical units of W/m^2. The differences in reflectivity of these three surfaces are quite dramatic and would make an interesting science project even for young students.

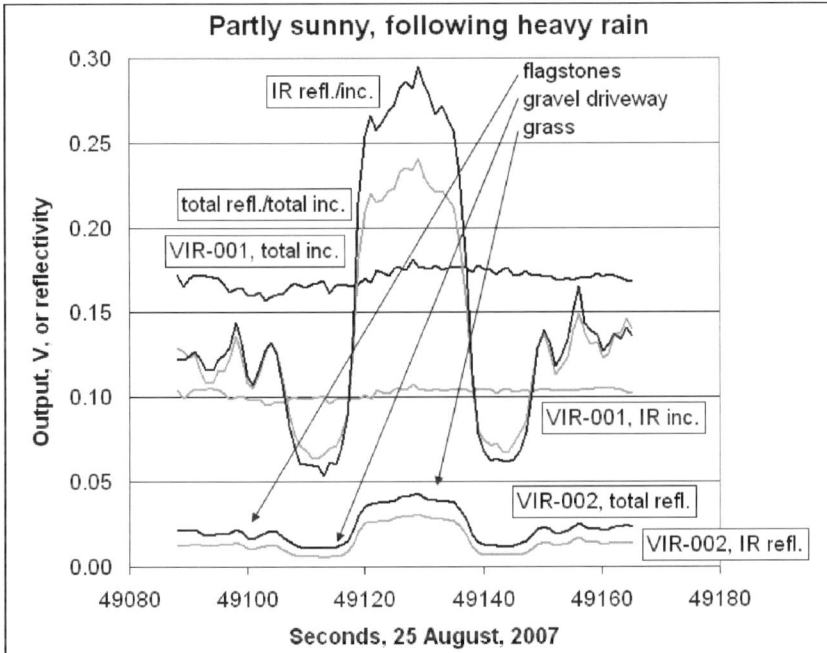

Figure 4.25. Reflectivity measurements for various surfaces.

Ideally, a reflectivity measurement should not depend on how far above a homogeneous surface the detectors are. However, this ideal is not attained in practice because radiometers are designed to respond to radiation from the entire hemisphere above the detector. This means that the downward-pointing detectors see light reflected not just from the surface directly below the detector, but from other surfaces going out to the horizon. There is a trade-off. The higher the detector is above the ground, the more representative of an average surface the reflected radiation will be, but the more likely the response is to be contaminated by reflections from other surfaces going out to the horizon.

It might be tempting to restrict the field of view of the downward-looking detectors to limit their response to just what is below them. However, this greatly complicates the relative calibration with the upward-looking instrument. Providing both sets of detectors with a field-of-view limiter means that the upward-looking instrument often may not receive direct sunlight, which is the major contributor to the reflected energy. So, the trade-off should be resolved in favor of unrestricted fields of view in both instruments, with the downward-viewing instrument close enough to the surface being measured to minimize contamination from other surfaces. In many cases, a distance of a meter or so above the surface should be sufficient to "see" a representative surface sample with minimum contamination.

It is important to position downward-viewing instruments as far as practical from their mounting platform—a person, tripod, or other structure—to avoid interference from the support structure itself. Also, in order to obtain consistent data for comparing various surfaces, it is important to mount the downward-viewing instrument always at the same height above the surface.

The reflectivity of the surfaces shown in Figure 4.25 ranges from a little more than 5% for gravel to a little less than 30% for grass in the near-IR. These values are typical of Earth's surfaces. Fresh snow can have a very high broadband reflectivity, >90%, and open oceans (as viewed from space) have a very low reflectivity, <5%. As is clearly evident from Figure 4.25, reflectivity is wavelength dependent, and there are some interesting differences in total and near-IR reflectivity among various surfaces which can easily be observed with the reflectometer described here.

An interesting extension of the broadband/near-IR instrument described here would be a multi-channel instrument including, for example, red and green LED detectors. If you imagine a grass lawn being green and lush in early summer and then parched and brownish later in the summer, it is easy to see that reflectivity at green wavelengths could serve as a measure of the health of vegetation.

5. Instrument Design Principles II: Sun Photometers

Chapter 5 discusses instruments for measuring radiation coming directly from the sun. It describes in detail a visible light sun photometer for measuring aerosols in the atmosphere, and a similar instrument that uses near-infrared radiation to measure water vapor in the atmosphere. It includes a discussion of how to use digital photography to study the solar aureole.

5.1 Measuring Aerosols

5.1.1 Designing an LED-Based Visible Light Sun Photometer

As described in Chapter 3, sun photometers are instruments that measure direct sunlight. The basic operating principle is that direct sunlight is scattered and absorbed as it passes through the atmosphere, and that the amount by which direct sunlight is diminished at Earth's surface depends on what is in the atmosphere. This principle is expressed by Beer's law. For an initial intensity of direct sunlight I_o at wavelength λ the intensity at the detector is

$$I_\lambda = I_{o,\lambda}\exp(-\alpha m_{air}) \qquad (5.1)$$

where α is the "optical thickness" of the atmosphere at wavelength λ and m_{air} is the relative air mass, equal to 1 when the sun is directly overhead and approximately equal to $1/\cos(z)$ where z is the solar zenith angle. (See Appendix 4 for details about calculating relative air mass.) The larger the optical thickness, the less light reaches the detector. One atmospheric constituent that is easily measured with a sun photometer is aerosols, which both absorb and scatter direct sunlight and which can be separated from other contributors to total optical thickness.

In order to approximate the theoretical requirement for monochromatic light, research-quality sun photometers use broadband detectors with "interference filters" to limit the incoming light to a range

71

of only a few nanometers.[1] Simple handheld sun photometers based on interference filters were first described by Frederick Volz [1974]. However, interference filters can be fragile, expensive, and subject to unpredictable degradation in their transmission properties. These are not desirable characteristics for an inexpensive instrument!

Sun photometers using light-emitting diodes (LEDs) as an alternative to interference filters were first described by Mims [1992]. LEDs are rugged, cheap, and very stable both electronically and optically. The same physical principles that allow LEDs to emit light when a current is passed through them allow them to generate a current when exposed to light of an appropriate wavelength. However, as described in Chapter 4, it requires expensive equipment to determine how a particular LED will respond to sunlight. An LED's spectral response is always different from its emission spectrum, with a peak response wavelength that lies below its peak emission wavelength. (Refer to Figure 4.14.)

For the application discussed in Chapter 4, an instrument to monitor photosynthetically active radiation, the wide spectral response of an LED detector was not a particular problem. However, it is a major design issue for sun photometers because of the monochromatic assumption required to apply Beer's law.

The first task in designing an LED-based sun photometer is to examine the spectral response properties of some potential LED detectors. The sun photometer described in this chapter has two channels in the visible part of the spectrum, one green and one red. There is no compelling scientific reason to choose these colors over others, but there are practical considerations. With an interference-filter-based instrument, the wavelengths can be set to whatever is desired. But with LEDs, the possibilities are limited by what is commercially available. Figure 5.1 shows the normalized measured spectral response of the two LEDs used in the sun photometer originally developed for the GLOBE Program in the late 1990s. Clearly, these detectors, with a spectral response spanning tens of nanometers rather than just a few nanometers, are not monochromatic detectors! However, they represent the best compromise choices available in the visible spectrum. The emerald green HLMP-D600 LED performs

---

[1] Interference filters consist of a substrate of glass or some other transparent material coated with thin layers of material with different refractive indices. The properties of these coatings determine which wavelengths of light can pass through the filter.

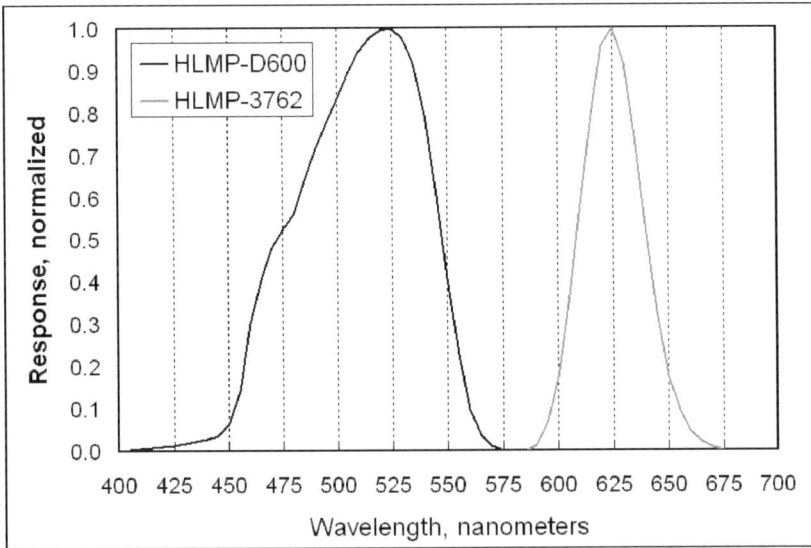

Figure 5.1. Normalized spectral response of green and red LEDs used in the two-channel visible light sun photometer.

better than other green LEDs, and some other red LEDs have much wider spectral response spectra than the HLMP-3762.[2]

Because of the violation of the monochromatic assumption for LEDs, the performance of a sun photometer using these detectors was studied extensively, and the results have been published in the peer-reviewed literature [Brooks and Mims, 2001]. Basically, the use of such devices requires definition of an "effective" single wavelength to which optical thickness will be assigned. For this instrument, the peak detector responses are at about 525 and 625 nm. However, because of the shape of the green LED response and the wavelength-dependent contribution of Rayleigh (molecular) scattering to total atmospheric optical thickness, the effective optical thickness wavelength for aerosols is about 505 nm for this instrument. The response of the red LED is narrower and more symmetric about its peak response. Also, the total contribution and variation with wavelength of Rayleigh scattering is much less at these wavelengths. As a result, the effective aerosol optical thickness wavelength is 625 nm, the same as the LED's peak response.

---

[2] As this book was being written, production of D600 and 3762 LEDs was being phased out. Alternatives will need to be evaluated on their merits, as these were in their initial selection.

The two-channel sun photo-meter described in this chapter is shown in Figure 5.2. There are two 5.5-mm (7/32") diameter holes in the top of the case. Light shines through these holes and the hole in the top alignment bracket. When the sun spot passing through the upper alignment bracket is centered around a dot on the lower bracket, the sunlight entering the case is centered on the detectors inside. These detectors are mounted on a printed circuit board located approximately under the observer's thumbs. The sun photo-meter's on/off switch is on the left in the photo. The knob on the right selects the channel to be displayed on the panel meter—temperature, green channel, or red channel.

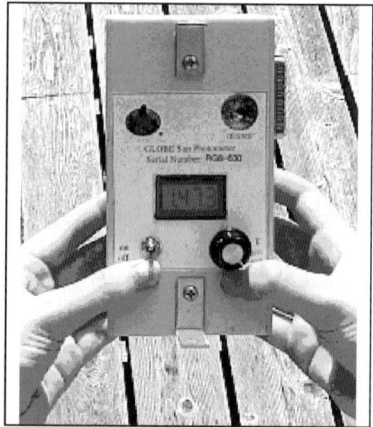

Figure 5.2. Two-channel LED-based visible light sun photometer.

The pc board containing the sun photometer's electronics is shown in Figure 5.3. The circuitry consists of a simple two-channel op-amp trans-impedance amplifier to convert the small current from the red and green LEDs (recall Table 4.2) to a usable DC output voltage, with gain resistors of 2.0 MΩ for the green channel and 5.6 MΩ for the red channel. These choices give a voltage output in the 1–2 V range for both channels. The digital panel meter will display voltages only up to 1.999 V, so it is important to limit the gain to no more than this value.

As is true for all the instruments discussed in this book, the smallest detail can have significant consequences. Recall that another assumption for applying Beer's law is that the detectors should "see" only direct sunlight. In the case design shown in Figure 5.2, the distance from the holes in the top of the case to the detectors is about 12.5 cm. This restricts the instrument's field of view to arcsin(0.55/12.5) = 2.5°. It is important to restrict the field of view, but the smaller the field of view, the harder it is to keep the sunlight centered on the LEDs. Also, if the holes through which sunlight enters the case are smaller than the LEDs them-selves (the LED housings are 5 mm in diameter), it will be very difficult for an observer to keep the sunlight centered on the LEDs. So, this case design is a compromise between theoretical requirements and practical considerations. Experience has proven that this design does not signi-ficantly impact the theoretical requirement for direct sunlight.

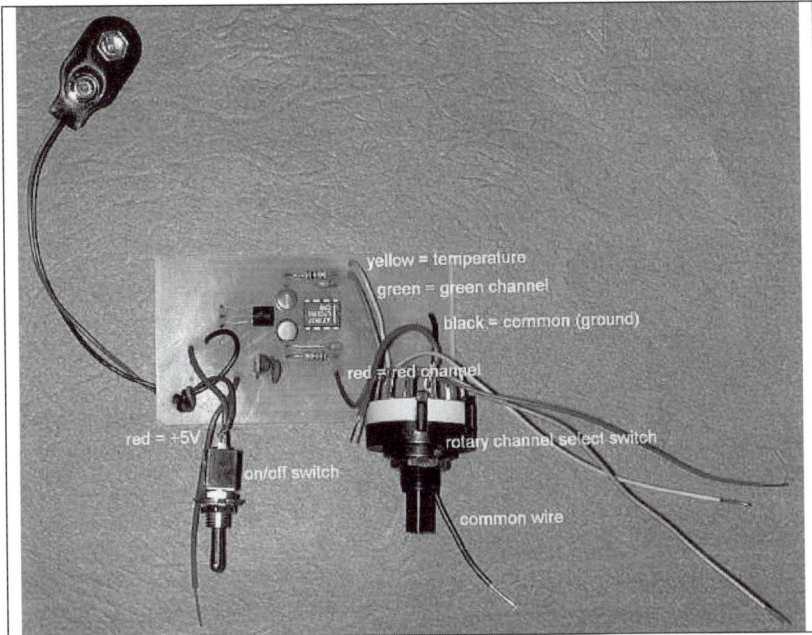

Figure 5.3 (see color plates). Printed circuit board assembly for two-channel sun photometer.

This instrument also includes an LM35DZ temperature sensor, visible just to the left of the two LEDs on the pc board shown in Figure 5.3. It is important to monitor the air temperature inside the case because the output of LEDs (and other photodetectors) depends on temperature. These instruments are designed and calibrated to be used over a narrow temperature range, roughly 20–25°C. If the temperature inside the case is significantly outside this range, measurements of optical thickness will not be accurate without a temperature correction. The overall size of the sun photometer case, which could certainly be made much smaller, helps to slow a rise or drop in air temperature inside the case when the instrument is taken outside to collect data.

5.1.2 Calibrating and Using a Sun Photometer

The output voltage from the transimpedance amplifier in this sun photometer is assumed to be proportional to the amount of sunlight reaching the detector. That is, $V \sim I$. Recalling Equations (3.2) and (3.5), and consolidating optical thickness into Rayleigh, α_R, and non-Rayleigh components, α_a, the equation for interpreting sun photometer measurements becomes:

$$V = V_o(R_o/R)^2 \exp[-(\alpha_a + \alpha_R p/p_o)m_{air}] \tag{5.2}$$

Solving for non-Rayleigh optical thickness, α_a,

$$\alpha_a = [ln(V_o(R_o/R)^2) - ln(V) - \alpha_R(p/p_o)m_{air}]/m_{air} \tag{5.3}$$

where ln is the natural (base e) logarithm, R is the Earth-sun distance at the time of a measurement, R_o is the average Earth-sun distance (1 astronomical unit), and p/p_o is the ratio of actual barometric pressure (station pressure) at the observing site to standard sea level barometric pressure (1013.25 mb).

As noted in Chapter 3, the wavelength-dependent Rayleigh scattering component can be calculated theoretically. For the LEDs used in this instrument, the effective optical thickness wavelengths have been shown to be about 505 nm for the green channel and 625 nm for the red channel [Brooks and Mims, 2001]. Using Equation (3.6) and the values in Table 3.1, the Rayleigh scattering contribution to total optical thickness at a standard sea level barometric pressure of 1013.25 mb is 0.138 for the green channel and 0.0159 for the red channel.

The contributions due to ozone (and perhaps other absorbing gases under some circumstances) can be separated from α_a after the fact, either by using climatological and latitude-dependent average ozone values, for example, or by using actual total column measurements for the time and place of the data collection. Satellite-based instruments such as the National Aeronautics and Space Administration's Total Ozone Mapping Spectrometer (TOMS) and instruments on the Aura spacecraft are sources of such data. For these two-channel sun photometers, a typical ozone contribution to the non-Rayleigh optical thickness is about 0.01 for the green (505 nm) channel and 0.03 for the red (625 nm) channel. Under almost all conditions under which this instrument can be used, aerosol optical thickness (AOT) is less than 1; a value of 2 would be a very extreme case. Very clean skies can have values of 0.05 or less, and in these circumstances, ozone optical thickness is a significant percentage of the total non-Rayleigh optical thickness.

Note that except for a few research sites, barometric pressure is invariably referenced to sea level when it is reported. Otherwise, it would not be possible to construct weather maps of low- and high-pressure systems. As a result of this convention, for example, the barometric pressure reported for Denver, Colorado, USA, "the mile high city," has the same range of values as it does in Philadelphia, Pennsylvania, near sea level.

But, for a standard atmosphere, the station pressure (the actual pressure) in Denver, at an elevation of about 1,600 m, is only about 830 mb. A conversion of "weather report" pressure to station pressure, accurate enough for any site that is likely to be accessible to readers of this book, is

$$p = p_o \cdot \exp(-0.119h - 0.0013h^2) \qquad\qquad (5.4)$$

where h is site elevation in km. If pressure is expressed in inches of mercury, as it almost always is in the U.S., the conversion is

$$p_{mb} = p_{in\,Hg} \cdot (1013.25/29.921) \qquad\qquad (5.5)$$

The remaining quantity in Equation (5.3), V_o, is the calibration constant for a particular instrument—the voltage output the instrument would produce if there were no atmosphere between the detector and the sun. This value will be different for every instrument because of variations in the electronics of the transimpedance amplifier and, more significantly, because the current output of LEDs, even from the same production batch, can vary significantly from sample to sample.

Obviously, it is not practical actually to measure the voltage outside the atmosphere. However, it is possible to infer what this voltage would be, based on measurements from the ground. This provides, in principle, a method of absolute calibration which is not available, for example, for a pyranometer, as discussed in Chapter 4.

If a sun photometer views the sun through various values of relative air mass, and the total atmospheric optical thickness does not change, then the logarithm of the instrument output voltage is proportional to relative air mass. By fitting a straight line through such data, the intersection of that line at the y-axis where m = 0 is the logarithm of the voltage the instrument would produce if there were no atmosphere. This is called a Langley plot calibration, named after Samuel P. Langley, the scientist who developed this technique in the early 20th century as part of efforts to determine "the solar constant of radiation," as described by Abbott and Fowle [1908].

It is not easy to obtain accurate Langley plot calibrations, because it is not easy to find observing sites at which total atmospheric optical thickness remains constant for several hours, as is required to collect data over a range of relative air mass values. High-elevation sites such as

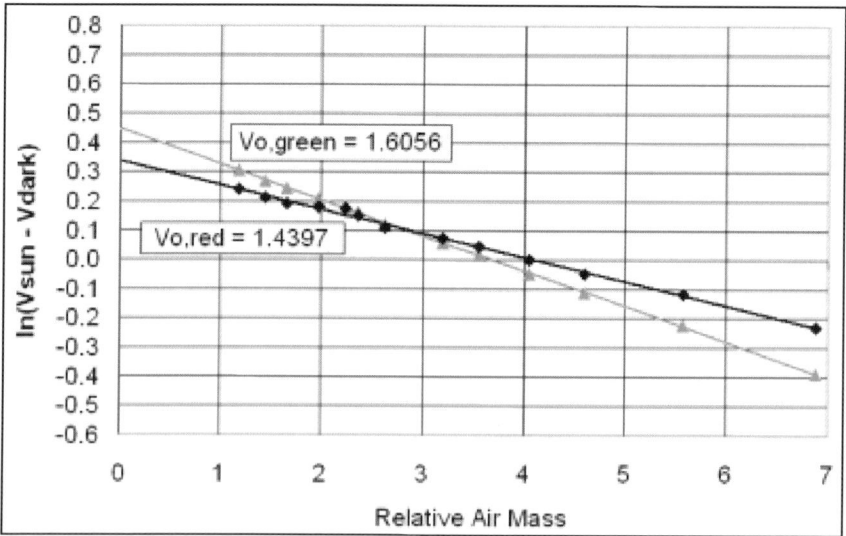

Figure 5.4. An example of a Langley plot sun photometer calibration, based on data collected at Mauna Loa Observatory, 29 May 2006.

Mauna Loa Observatory (MLO) in Hawaii are favored for such work because of the clean skies and stable meteorology.[3] Figure 5.4 shows a Langley plot calibration performed on a two-channel LED sun photometer at MLO on June 21, 2000 [unpublished data by Brooks and Mims, 2006].

Taking the green channel as an example, the logarithm of the voltage extrapolated to $m_{air} = 0$ is 0.44713. This gives a voltage of 1.56382. The value V_o should be normalized to an average Earth-sun distance (1 AU). On May 29, 2006, the Earth-sun distance was 1.013286 AU. The intensity of radiation from the sun, and hence the voltage, varies as the square of the Earth-sun distance, so

$$V_o = (1.56382)(1.01631)^2 = 1.60565 \qquad (5.6)$$

That is, if the instrument were 1 AU from the sun, the voltage would be larger than it was on June 21.

---

[3] Because of increasing amounts of dust and air pollution crossing the Pacific Ocean from Asia, skies above MLO are much less clean than they used to be. Also, MLO weather can sometimes be unpredictable, just as at every other observing site!

Of course, it is expensive to travel to MLO[4] and this is not a practical way to calibrate large numbers of sun photometers. Sun photometers used in the field can be calibrated by comparing their output side-by-side with a reference instrument on which a Langley plot calibration has been performed. This is called a transfer calibration, and is has been performed on literally hundreds of the sun photometers described in this chapter.

An alternate source of transfer calibrations is aerosol optical thickness data from the National Aeronautics and Space Administration's Aerosol Robotic Network (AERONET) [Holben *et al.*, 1998]. For anyone fortunate enough to live near one of these instruments, an AERONET transfer calibration has the advantage of tying the performance of an instrument directly to a widely accepted source of AOT data. Although AERONET instruments are very expensive (>$20,000), they still have the same calibration requirements. The AERONET program also conducts Langley calibrations at MLO for its reference instruments and calibrates its field instruments against these very valuable standards.

Like all photodetectors, the LEDs used in this sun photometer have a temperature dependence. This quantity cannot be determined from manufacturers' specifications but must be measured under realistic conditions. As was true for the pyranometer discussed in Chapter 4, the problem is that there is no easy way to measure the temperature of the detector in the LED. The basic approach is to use a reference instrument—one that has been calibrated at MLO, as described above—to calibrate a test instrument. Initially, both instruments are placed side by side to stabilize their detector temperatures in an environment where the ambient air temperature is around 22°C. After the air temperatures inside the cases have stabilized to the same value, data are collected from both instruments, and values of V_o for the green and red channels are calculated so that the derived optical thickness values are the same. Then, this process is repeated with the test instrument stabilized to a series of warmer or colder temperatures. For warmer temperatures, the test instrument can be placed in a sunny location. For colder temperatures, it can be refrigerated or, in the winter, placed outside. This is a process that requires good weather, with access to an unobstructed view of the sun over several hours.

Prior to each set of measurements it is necessary to assume that the air temperature in the case is the same as the detector temperatures. The data must be collected quickly, using a data logger, because once the

---

[4] MLO is a remote high-elevation research site accessible only by a drive of several hours along a poorly maintained road. The facilities there are extremely limited. This is definitely not the Hawaii described in travel brochures!

sun photometer is aligned with the sun, the LEDs begin to heat from the solar input and the assumption that the case temperature is the same as the detector temperature is no longer valid.

Figure 5.5 shows results from a temperature sensitivity study conducted during 2007. The initial measurements were done in the summer, with the test instrument stabilized at several different temperatures warmer than the reference instrument. From these data, an equation for correcting the V_o values as a function of temperature was derived. Then, during the winter, additional measurements were made with the same reference instrument and a different test instrument subjected to colder temperatures.

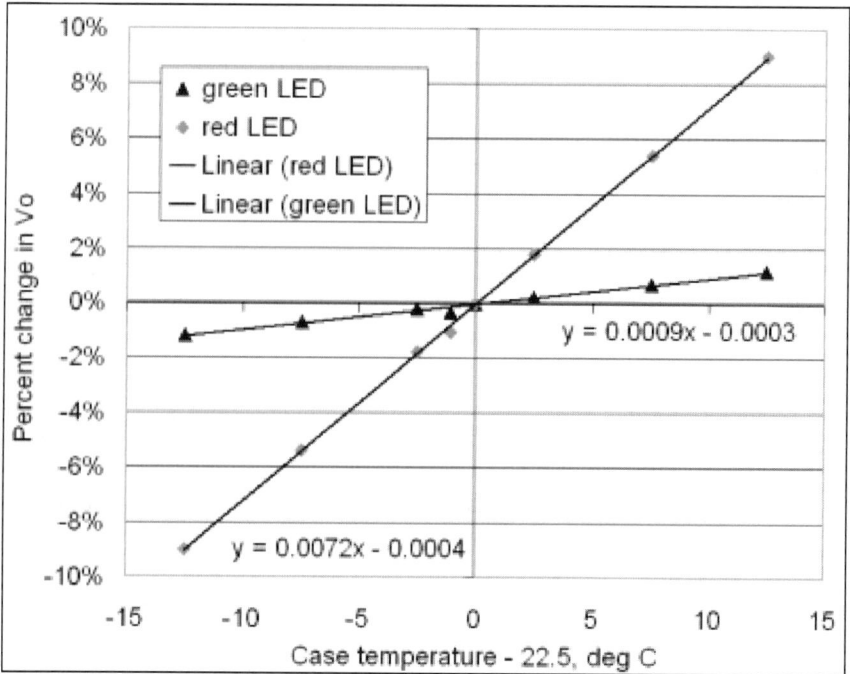

Figure 5.5. Temperature compensation curves for LED sun photometer calibration constants.

The question this experiment was designed to answer is, "Will the temperature correction equation derived for one instrument subjected to warm temperatures also apply to a different instrument subjected to cold temperatures?" Even though all the sun photometers built as described in this chapter use the same kinds of LEDs, the current outputs of LEDs vary significantly from sample to sample, which leads to a significant variation in the V_o among instruments. However, it is reasonable to hope that the percent change in the V_o values should be the same for different instruments.

The data shown in Figure 5.5 demonstrate that the answer to this question is "Yes." The summer data, with test temperatures above 22.5°C, and the winter data, with test temperatures below 22.5°C fall along the same straight line, even though the two data sets were collected at different times of the year and were obtained from two different instruments. These data show that the red LED is much more temperature sensitive than the green LED, a fact that has always been qualitatively obvious to users of these instruments.

5.1.3 Applications

Appropriate weather conditions and considerable care are required to obtain reliable aerosol optical thickness values with a handheld sun photometer. The sky does not need to be cloud-free, but the observer must have an unobstructed view of the sun. Ideally, the instrument should be stabilized at a temperature around 22.5°C, where it was calibrated initially. A typical procedure is to take three sets of green/red channel measurements as quickly as possible, to minimize the effects of heating on the output voltages. Even though Figure 5.5 gives temperature corrections, these can be applied only when the instrument and its detectors are at a stabilized temperature, not while the detectors are exposed to sunlight and their temperatures are changing faster than the air temperature inside the case.

With a little practice, a set of sun photometer measurements can be collected manually in no more than 2 min. The temperature inside the case should be recorded immediately before and after the voltages are recorded. The time for each set of measurements should also be recorded to the nearest 10 s or so—accurate times are needed to calculate the position of the sun and, from that, the relative air mass. The barometric pressure at the observer's site must also be recorded. This value can easily be obtained from online sources before or after recording the sun photometer data. Barometric pressure changes relatively slowly, and the calculations are only weakly dependent on pressure, so the value for the nearest hour is sufficient, but that value must be corrected to station pressure, using Equation (5.4).

These visible-light sun photometers have been used around the world for both educational and scientific purposes [*e.g.*, Brooks *et al.*, 2003; Boersma and de Vroom, 2006]. Figure 5.6 shows aerosol optical thickness data recorded at an elementary school in rural Arkansas. In temperate climates, AOT values are typically higher in the summer than they are in the winter. However, these data illustrate a common problem with making these measurements in schools: during the summer, there is often no one available to collect data! Using two instruments provides valuable quality control information about the performance of the instruments. When measurements are made under the same conditions, the agreement between these instruments is excellent.

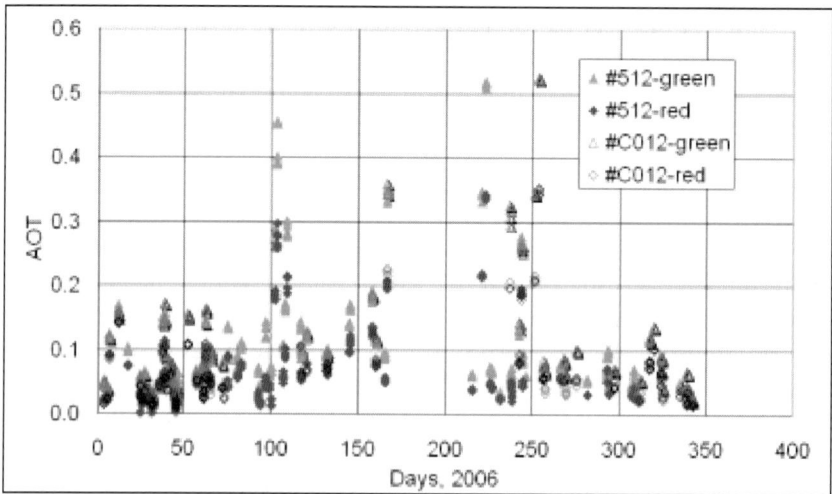

Figure 5.6 (see color plates). Aerosol optical thickness data from two sun photometers at a rural school in Arkansas. Data provided by Wade Geery. Latitude 36.1972° N, longitude 92.2688° W.

5.2 Measuring Water Vapor

5.2.1 Designing a Near-Infrared Sun Photometer for Detecting Water Vapor

A water vapor sun photometer is physically identical to the visible-light sun photometer shown in Figure 5.2 [Brooks *et al.*, 2003]. Only the detectors are different, to take advantage of water vapor absorption in the near-infrared.

Recall that water vapor molecules absorb solar radiation at specific wavelengths. (You can see several water vapor absorption "holes" in Figure 3.1) As a result, one way to determine total atmospheric water vapor (precipitable water, or PW) is to measure the ratio of directly transmitted sunlight at two wavelengths, one inside a water vapor absorption band and one outside the band. As long as the transmission of sunlight at each of these two wavelengths is not affected differentially by some other atmospheric constituent, this ratio can be related to PW.

Figure 5.7 shows, first of all, the spectral variation of insolation for a standard atmosphere across the near-IR part of the solar spectrum, with the sun directly overhead. Note the water vapor absorption band centered around about 940 nm and a smaller band around 820 nm.

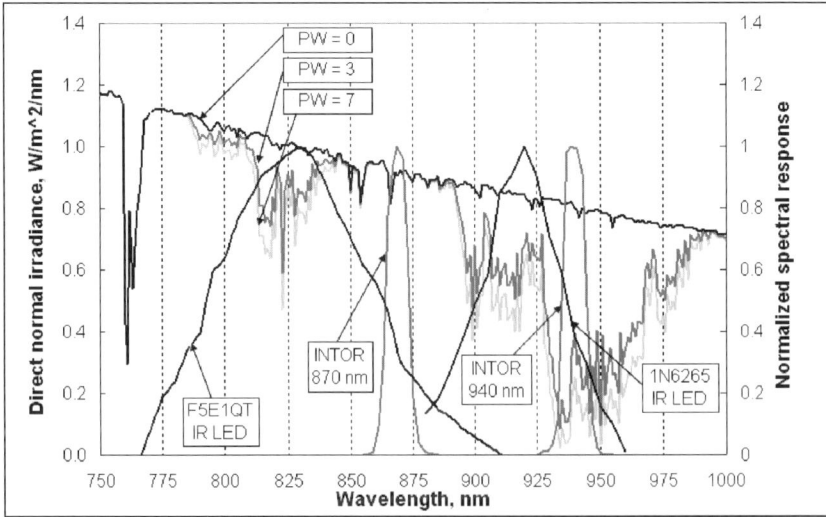

Figure 5.7. Water vapor absorption bands in the near-IR and normalized response of possible near-IR detectors [Irradiance data from SMARTS2 model, Gueymard, 1995].

Figure 5.7 also shows the normalized spectral response of two possible near-IR LED detectors. In his original design for a sun photo-meter to measure PW, Mims [1992] suggested the use of two IR LEDs. However, Figure 5.7 illustrates some of the problems arising from the use of detectors whose response cannot be tailored for a specific application. The 870-nm LED response (870 specifies its peak emission wavelength) overlaps a small water vapor absorption band around 825 nm. The 940-nm LED response is well below the peak of the water vapor absorption band centered around 940 nm.

An even more serious problem plagues the 940 nm LED, although it is not evident from Figure 5.7. Its output depends so strongly on temperature that it is not a reasonable choice for this instrument.

Fortunately, advances in filter manufacturing technology have made possible some compromises for this instrument, although at a slightly higher cost than if it used only LED detectors. A typical LED costs roughly $1. A 940-nm filtered photodiode in a similar size housing costs about $20. Its spectral response is also shown in Figure 5.7.

The decision to replace the 870 nm LED with a filtered photodiode also seems to make sense in order avoid the water vapor absorption around 825 nm. The response of an 870-nm filtered photodiode is also shown in Figure 5.7. However, during the development of this instrument, tests of samples with one and two filtered photodiodes showed that the slightly less expensive instrument with the 870-nm LED and a 940-nm filtered photo-diode performed about the same as the instrument with no LEDs.

As noted at the beginning of this section, the water vapor sun photometer is physically identical to the two-channel visible-light sun photometer discussed earlier in this chapter, except for its detectors. Different detectors require different gain resistors. Whereas gain resistors for the green and red LED detector channels are typically 2.0 MΩ and 5.6 MΩ, gain resistor values 10 to 20 times smaller are sufficient for this instrument to give an output in the 1–2 V range in full sunlight, depending on whether near-IR LEDs or filtered photodiodes are used.

5.2.2 Calibrating and Interpreting a Water Vapor Sun Photometer

Atmosperhic water vapor is usually measured in units of centimeters of water (cm H_2O). Consider the vertical column of air directly above an observer holding a cylindrical cup. If all the water vapor in that column could be condensed into the cup, the depth of the water in the cup would be the total column precipitable water vapor (PW). The accepted standard for presenting measured PW values is to convert them to the corresponding values of cm H_2O in the overhead column—at a relative air mass of 1. An approximate conversion when the solar zenith angle is not too large is:

$$PW_{m=1} = PW_m \cdot \cos(z) \text{ cm } H_2O \tag{5.7}$$

There are no easy ways to obtain direct measurements of total column water vapor for the purpose of calibrating near-IR sun photo-meters. Balloon-borne instruments that rise into the stratosphere, beyond essentially all atmospheric water vapor, can be used to measure total column water vapor as well as vertical profiles. However, these measurements are spatially sparse and sporadic.

Reitan [1963] developed an approximate relationship between PW and Celsius dewpoint temperature, T_d, based on meteorological para-meters at Earth's surface. Converting the original formulation from Fahren-heit to Celsius degrees, Reitan's equation is:

$$PW_{Reitan} = \exp(0.1102 + 0.0614T_d) \tag{5.8}$$

    A somewhat more sophisticated model replaces Reitan's constant coefficients with coefficients that depend on season and (northern hemisphere) latitude [Smith, 1966]:

$$PW_{Smith} = \exp[0.1133 - ln\,(C_{Smith} + 1) + 0.0393(9T_d/5 + 32)] \qquad (5.9)$$

where the coefficient C is chosen from the value in Table 5.1. (Smith's model is also based on temperature in degrees Fahrenheit, which explains the conversion from Celsius to Fahrenheit in the equation.)

Table 5.1. Seasonal and latitude-dependent coefficients, C_{Smith}, for calculating precipitable water vapor from dewpoint temperature.

Latitude	Winter	Spring	Summer	Fall
0–10	3.37	2.85	2.80	2.64
10–20	2.99	3.02	2.70	2.93
20–30	3.60	3.00	2.98	2.93
30-40	3.04	3.11	2.92	2.94
40–50	2.70	2.95	2.77	2.71
50–60	2.52	3.07	2.67	2.93
60–70	1.76	2.69	2.61	2.61
70–80	1.60	1.67,	2.24	2.63
80–90	1.11	1.44	1.94	2.02

    Dewpoint temperature is available from some online weather sources, including National Weather Service current weather reports in the United States. One formulation for calculating dewpoint temperature from air temperature and relative humidity, in degrees Celsius [NOAA, online 2007], is:

$$e_s = 6.11 \times 10^{[7.5T/(237.7 + T)]}$$
$$T_d = [237.7 \log_{10}(e_s \times RH/611)]/[7.5 - \log_{10}(e_s \times RH/611)] \qquad (5.10)$$

where e_s is the saturated vapor pressure of the atmosphere, T is the air temperature in degrees Celsius, and RH is relative humidity, expressed as a percent.

    The coefficients in equations (5.8) and (5.9) provide a statistical best fit to data but do not imply that the PW value is accurate to this many significant digits. Indeed, there are often weather conditions, such as the rapid passage of a front, during which both Reitan's and Smith's equations will perform poorly in predicting PW. Figure 5.8 shows PW values at Millersville, Pennsylvania, during 2007, calculated from signals received from the Global Positioning Satellite (GPS) system (GPS-MET) [Gutman and Holub, 2000; Tregoning *et al.*, 1998]. Water vapor delays the transmission of radio signals through the atmosphere and this delay can be related to total column water vapor.

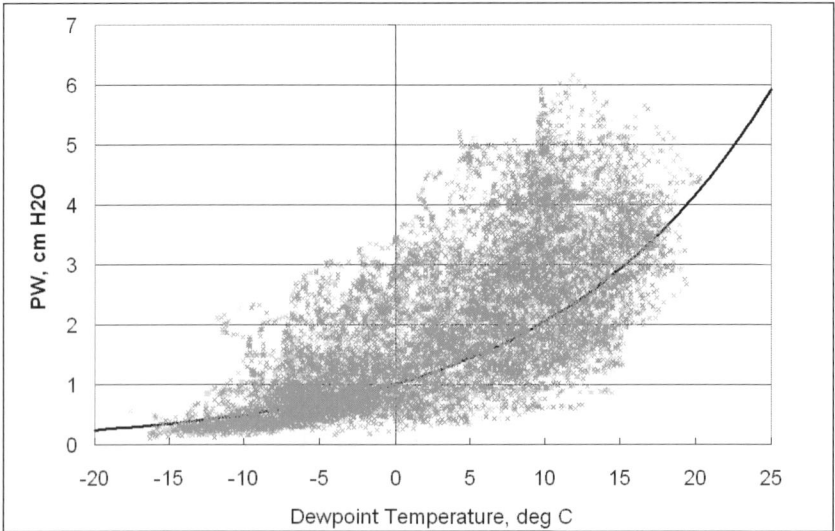

Figure 5.8. GPS-MET-based PW vs. dewpoint temperature, measured over Millersville, Pennsylvania, 2007.

The solid line in Figure 5.8 represents PW as a function of dewpoint temperature, calculated using Equation (5.9) with the average of the eight 30–40° and 40–50° latitude coefficients given in Table 5.1. (Millersville is at 40° N latitude.) Based on these data, it is clear that it is not really possible to determine total column water vapor accurately from conditions on the ground. After all, if this were possible, there would be no need for a water vapor sun photometer!

It is also possible to use sun photometer measurements along with sophisticated models of the transmission properties of the atmosphere to determine water vapor. This has been done with the CIMEL instruments in the AERONET sun photometer network. However, these calculations cannot be duplicated with the instruments described in this book. For the instrument described here, voltage outputs produced by the two near-IR channels can be represented as:

$$V_1 = V_{o,1}\exp(-m_{air}\alpha_{\lambda,1}) \tag{5.11a}$$

$$V_2 = V_{o,2}\exp(-m_{air}\alpha_{\lambda,2})T_{WV} \tag{5.11b}$$

where T_{WV} represents the reduction in direct sunlight due to the presence of water vapor. One well-known formulation for T_{WV}, dating back to the 1960s [Gates and Harrop, 1963], is:

$$T_{WV} = a - b(m_{air} \cdot PW)^\beta \qquad (5.12)$$

where PW is the precipitable water vapor. By convention, PW refers to the amount of water vapor directly over the viewer, so multiplying this value by m_{air} gives the amount of water vapor the instrument actually "sees" when it is pointed at the sun. From Equations (5.11) and (5.12),

$$ln(V_2/V_1) = ln(V_{o,2}/V_{o,1}) + m \cdot (\alpha_{\lambda,1} - \alpha_{\lambda,2}) + [a - b \cdot (m_{air} \cdot PW)\beta] \qquad (5.13)$$

where a, b, and β are derived from the spectral response characteristics of the channel 2 detector and models of sunlight transmission through the atmosphere.

For a particular instrument configuration, Equation (5.13) can be simplified as:

$$ln(V_2/V_1) = A + B \cdot [Cm_{air}\alpha_{500nm} - (m_{air} \cdot PW)^\beta] \qquad (5.14)$$

The optical thickness at the wavelength of each of the two channels is not known, nor are these values easily obtained. Hence, the optical thickness value at 500 nm times the constant C is substituted for the difference. AOT at 500 nm is chosen because this value is widely used as a way to characterize atmospheric aerosols.

In practice, it has been found that the constants B, C, and β are fixed for a particular instrument configuration, and that only the value of A needs to be determined for a particular instrument. This value depends on the output from the detectors, which varies from instrument to instrument.

Water vapor reference instruments can be calibrated against AERONET or GPS-MET data, as mentioned above. For large numbers of "field" instruments, a transfer calibration approach is used, as was done for the visible-light sun photometers described earlier in this chapter. Given a value of A for a reference instrument, the value for a test instrument is calculated by making simultaneous measurements with both instruments:

$$ln(V_2/V_1)_{ref} - A_{ref} = B[Cm_{air}\alpha_{500nm} - (m_{air} \cdot PW)^\beta] = ln(V_2/V_1)_{test} - A_{test}$$
$$(5.15a)$$

Solving for A_{test}:

$$A_{test} = ln\,(V_2/V_1)_{test} - ln\,(V_2/V_1)_{ref} + A_{ref} \qquad\qquad (5.15b)$$

This work on calibration techniques for simple water vapor sun photometers has been described in the peer reviewed literature by Brooks *et al.* [2007].

Figure 5.9 shows how GPS-MET data for water vapor can be used to calibrate the water vapor instrument described here. There are no GPS sites within a few kilometers of Philadelphia, Pennsylvania, where this work was done, so this is not an ideal calibration site. However, there are two sites several tens of km from Philadelphia, one about 160 km to the west and another about 80 km to the southwest. As shown by the heavy pair of lines in Figure 5.9, the sites give similar, but by no means identical, PW values.

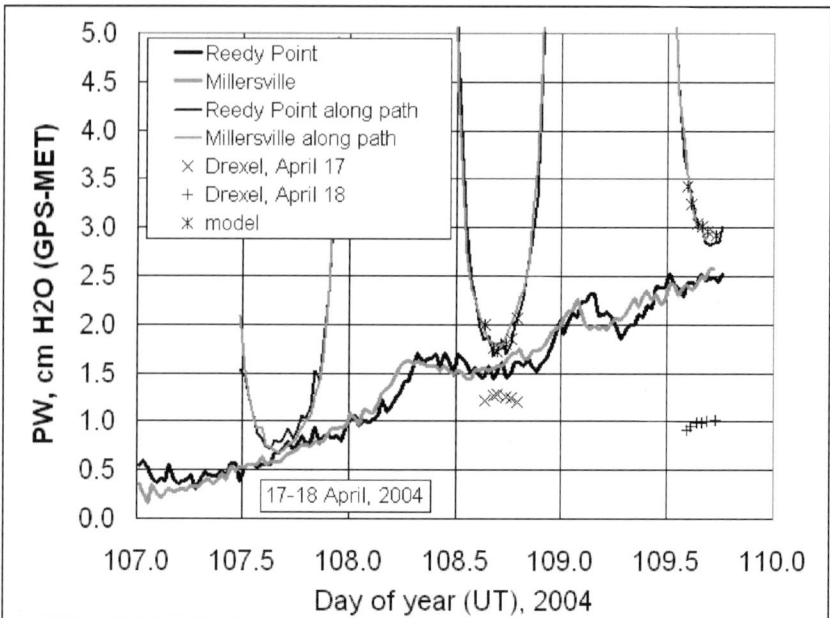

Figure 5.9. Calibrating a near-IR water vapor instrument using GPS-MET data.

Remember that processed PW data are always given relative to the atmosphere directly over the observer (a relative air mass of 1). In order to compare these values with the ratio of the voltage outputs from the two IR detectors in a sun photometer, the overhead PW values are converted to their values along the relative air mass m viewed by the sun photometer:

$$PW_{path} = m_{air} \cdot PW_{overhead} \qquad (5.16)$$

These converted values are shown as the thin pairs of lines in Figure 5.9. The IR detector ratios, V_2/V_1, are the lower sets of ×'s and +'s on Figure 5.9. With a model as described above, these ratios can be related directly to the PW values along the path between the observer and the sun. In Figure 5.9, the data with the sun photometer are modeled against the average of PW values from the two GPS sites and the results are shown as the ×'s that lie along the thin lines. If such a model can be applied over an appropriate range of PW values, as obtained by looking at the sun over a wide range of relative air masses, then the instrument can be considered to be calibrated against the GPS data.

5.2.3 Applications

Figure 5.10 shows PW values from data supplied by a secondary school in Puerto Rico. Sun photometer WV-116 was calibrated against a GPS-MET site about 10 km from this school in 2004 and has been used regularly since then to collect PW data. The calibration coefficients in Table 5.2 correspond to those given in Equation (5.15). The assumed aerosol optical thickness value of 0.10 at α_{500nm} can be replaced by measured values if they are available, as they often are from sun photometer measurements made at this school. (The 505 nm green channel AOT value can be used for this purpose.)

Table 5.2. Calibration coefficients for WV-116.

WV-116 coefficient	Value
A	0.8489
B	0.4700
C	0.20
B	0.65
α_{500nm}	0.10

Considering the differences between sun photometer measurements and GPS-MET measurements, the agreement is remarkable. Not only are GPS-MET PW values averaged over 30 min, but they are also averaged over a large part of the sky, depending on where the GPS satellites are at the time. Sun photometer measurements, on the other hand, represent instantaneous measurements of PW along the line of site from the observer to the sun. The agreement implies that PW does not vary significantly on the order of tens of kilometers around the observer, and that it changes rather slowly during the day.

Like aerosols, water vapor tends to be higher in the summer. The GPS-MET site was out of service during most of the second half of 2005, and the handheld instrument can be used to partially fill those data gaps—of course, only during the daytime when there is an unobstructed view of the sun for collecting data.

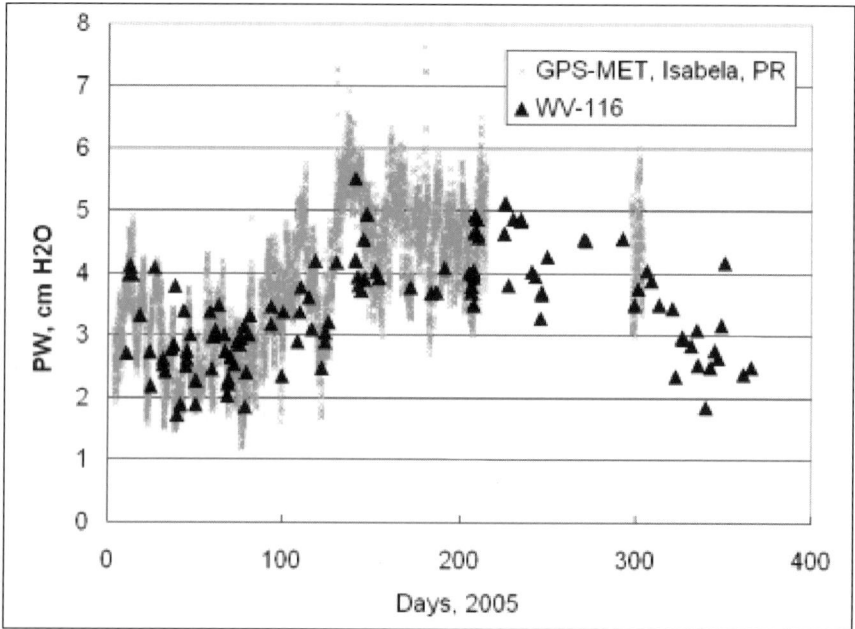

Figure 5.10. 2005 water vapor data from a GPS-MET site at Isabela, Puerto Rico, and Ramey School, Aguadila, Puerto Rico [WV-116 data courtesy of Richard Roettger].

5.3 A Different Way of Looking at Scattered Light in the Atmosphere

5.3.1 Designing a Fixture for Measuring the Solar Aureole

The instruments in this and the previous chapters have used various inexpensive detectors with simple electronics to measure quantitatively the total or direct solar radiation reaching Earth's surface—either broadband radiation or spectrally selective radiation. This section presents an entirely different approach to atmospheric measurements.

The size of the solar aureole—the whitish circular ring around the sun—is related to the amount of sunlight scattering in the atmosphere. That is, it is related at least qualitatively to the amount of aerosols and water vapor in the atmosphere. The larger the aureole, the hazier the atmosphere. Very clean skies at high elevation sites produce very small aureoles. Hazy summer days in temperate and tropical climates produce very large aureoles.

Although solar aureoles are easy to observe, until recently it has not been easy to analyze them quantitatively. However, digital photography provides a very simple and direct way to analyze aureoles. This requires only:

1. A digital camera with manual controls for setting focus, exposure, and f-stop.
2. Image-processing software for measuring the brightness of the sky along a line from the edge of the sun to some other point in the sky.

Currently, many digital cameras, even relatively inexpensive ones, include full manual control capabilities. The image resolution is not critical for this application—5 megapixel images are common even for low-end cameras and that is more than sufficient for this purpose. A large optical zoom capability is also not needed. In fact, none of the many sophisticated capabilities commonly found in digital cameras are needed for this project.

As for the software required to analyze aureole images, the *ImageJ* program, developed by the United States National Institutes of Health, is perfect for this task, and it is free![5]

To photograph the sky around the sun, it is necessary to block the solar disk itself. Otherwise, the light from the sun will "wipe out" the rest of the image. In principle, you can use the corner of a building or some other object. However, there is an important consideration when using digital cameras to photograph the solar aureole. Unlike film cameras, digital cameras capture images without a physical shutter between the lens and the electronic equivalent of a film frame. This means that the lens is, essentially, always open when the camera is turned on. (This is true even if the camera includes an optical viewfinder.[6]) With this arrangement, pointing the camera in the vicinity of the sun is definitely *not* a good idea.

---

[5] http://rsb.info.nih.gov/ij/
[6] A single lens reflex (SLR) film camera has a mirror between the lens and the film that reflects what the camera sees through the viewfinder. When the shutter is pressed, the mirror rotates out of the way and the film is exposed.

It would be even a worse idea to point the camera at the sun by looking through an optical viewfinder! Sunlight is extremely intense and focusing sunlight through a lens can permanently damage your camera's light detecting surface, not to mention your eyes! The instruction manuals of some digital cameras include specific warnings against pointing the camera at the sun.

The solution is to build a simple fixture that blocks light coming directly from the solar disk. Figure 5.11 shows such a fixture, mounted on a standard camera tripod. The mounting bracket is made from a 66-cm long piece of 1/4" × 1" aluminum bar stock,[7] available at hardware and building supply stores. The camera is mounted at one end of the bar with a 1/4"-20 screw, using the standard threaded tripod mount found on cameras. Another hole further down the bar is tapped for a 1/4"-20 screw, to attach it to the tripod. In between, a small hinge supports a sheet of thin aluminum approximately the same size as the camera body. This protects the camera from the sun while the setup is being positioned.

A penny is fastened with epoxy to a short piece of #14 copper wire and attached to the end of the bar with a #8 machine screw and washer. When the camera shown in Figure 5.11 is turned on with its lens extended, the end of the lens is about 61 cm from the penny. The penny should just cover the solar disk, leaving only the solar aureole and the sky around it. The shadow cast by the penny should cover the camera's lens. The precise positioning of the penny needs to be determined by trial and error, as described below in Section 5.3.2. Finally, the entire assembly visible from the camera should be spray-painted flat black to minimize light reflections.

To position the penny relative to the camera lens, paste a white paper label to the side of the aluminum sheet facing away from the camera. With the camera turned off, flip the aluminum sheet down out of the way. Adjust the tripod so the shadow from the penny is centered over the camera's lens cover. Then flip the sheet into the "up" position and draw a circle around the shadow cast by the penny.

To photograph the aureole, position the aluminum sheet in the "up" position to keep sunlight off the camera face. Turn the camera on, manually set its shortest exposure and smallest available aperture (largest f-stop), and focus it at infinity. Then adjust the tripod so the sun's shadow is centered on the circular target. When the position is set, flip the aluminum

---

[7] I apologize for the non-metric units, but this is the way the material is sold in the United States. The same comment applies to the reference to the 1/4"-20 screw thread that is standard in camera tripods—the 1/4" refers to the screw body diameter and the "20" refers to threads per inch.

Figure 5.11. Tripod fixture for photographing the solar aureole.

sheet down out of the way and take the picture, as shown in the inset photo. The camera shown in Figure 5.11 is a Canon PowerShot A530 set at 1/1,600 s and f-5.6. The camera retains these settings once they are set in manual mode, but the focus must still be set manually each time the camera is used.

Only a few seconds are required to align the penny's shadow and take a photo, during which time the sun will not have moved significantly. The result is an image that blocks direct light from the sun, but gives a clear view of the surrounding aureole.

5.3.2 Interpreting Aureole Images

Figure 5.12 shows *ImageJ* processing for an aureole image taken on a clear winter day. (The aerosol optical thickness values at 505 nm and 625 nm, obtained with measurements from the sun photometer described earlier in this chapter, were about 0.08 and 0.06.) A straight line has been drawn from the center of the disk covering the sun to the upper right hand corner of the picture. The graph shows the brightness along this line, analyzed as an 8-bit grayscale value between 0 and 255. The data plotted on the graph can be saved as a text file so they can be imported into a spreadsheet for further analysis.

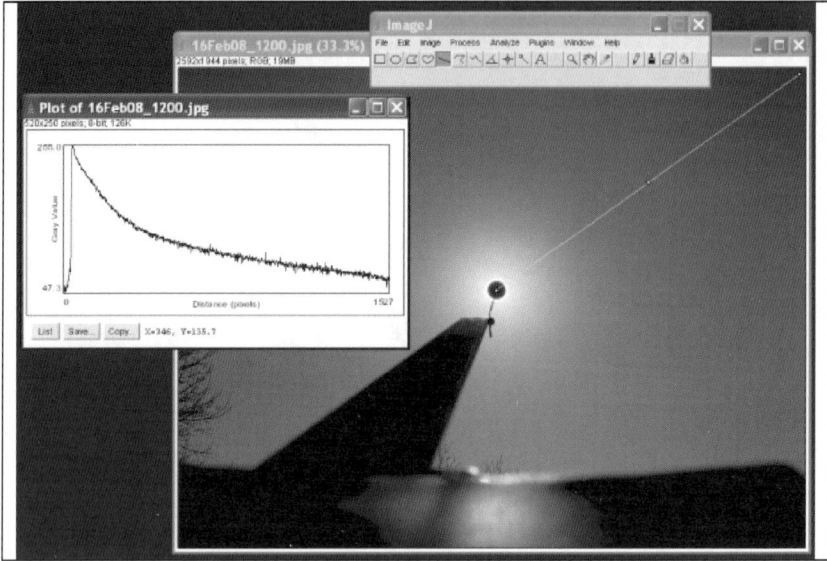

Figure 5.12. Aureole photo, 12:00 EST, 16 February 2008, processed by *ImageJ* software.

The density plot shown in Figure 5.12 will help to determine when the penny is correctly positioned to cover only the solar disk. The distance from the lens to the penny is right when the density plot on a very clear day reaches a value of 255—the maximum brightness the image will record—just at the edge of the disk. If there is a wider band of 255 values around the shadow, then the disk is too far away (or too small). If the density doesn't reach 255, then the disk is too close (or too large). For your camera, you may need to use a disk of a different size placed at a different distance from the lens.

The quantitative interpretation of aureoles, and particularly their relationship to aerosol optical thickness, remains an open and interesting question. However, it is easy to observe qualitatively some pronounced changes in the solar aureole that depend on atmospheric conditions.

Figure 5.13 shows density plots for two aureoles taken 6 days apart. The top plot was taken during a day on which some early morning cirrus clouds appeared to disappear by mid-day. However, the aureole photo showed what appeared to be the remains of some very thin and wispy cirrus clouds around the sun. These cloud remnants are clearly visible in the original color image, but may be hard to see in the small grayscale image included in Figure 5.13. In any case, the size of the aureole is clearly larger than it is in the lower photo, which is associated with the bottom density plot. This photo was taken following the very

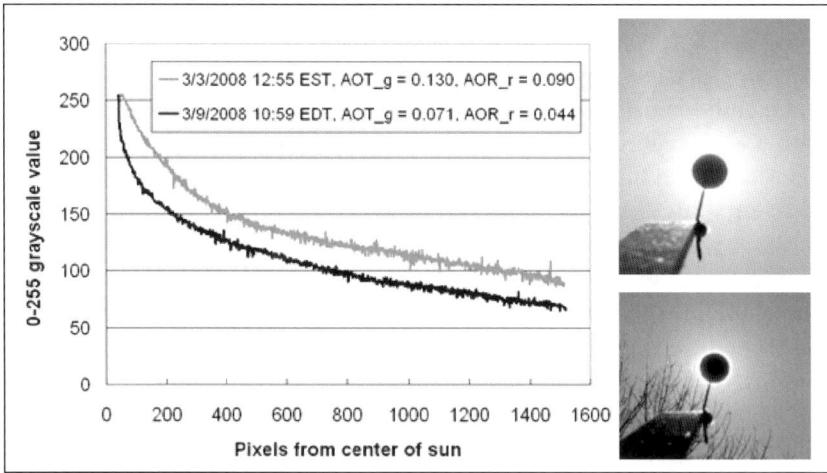

Figure 5.13. Aureole density plots for two sky conditions.

dramatic passage of a cold front. The day before the photo was taken, air temperature dropped from 18°C around midday to 1°C by late afternoon, with high winds.

Aerosol optical thickness (AOT) values at 505 nm (green) and 625 nm (red) are shown on the graph. If cirrus clouds really were present during the measurement on March 3, then the AOT values are suspect. Nonetheless, these are conditions under which almost any observer would have been willing to collect sun photometer data. The fact that decisions about whether an observer has an "unobstructed" view of the sun, as required for sun photometry, are difficult for human observers (and for computer-based cloud detection algorithms, too), reinforces the value of this kind of photography.

The rapid decrease in the aureole density seen on March 9 is observed at low elevations (this site is at about 120 m) only under conditions of clean skies with low AOT values. Very high elevation sites can have very small aureoles.

For the purpose of comparing and analyzing aureole images, it is important always to use the same camera, because different cameras process light differently and have a different field of view, and always to use the same exposure and f-stop. (Both images in Figure 5.13 were taken at 1/1,600 s, f-5.6.) If the camera is allowed to use its automatic exposure settings, you will not be able to compare images taken at different times under different sky conditions. If you can afford it, it is a good idea to devote a camera exclusively to this project, so it can be left permanently mounted on the fixture.

It is worth repeating here the warning given earlier about aureole photography. If you are careless about focusing direct sunlight through your camera's lens, or perhaps even if you are careful, it is *possible* to permanently damage your camera. So, undertake this project with that risk in mind. If you purchase a camera specifically for this project, get the least expensive model that will provide manual settings for focus, exposure time, and f-stop.

A simpler version of this project is to take photos of the sky which do not include the sun. A series of photos of the same scene, looking north, for example, taken daily near solar noon provide a very interesting record of sky conditions, including haze. You should include the horizon in such photos. If you are fortunate enough to be able to see some distant objects, such as tall buildings or mountains, from your observing site, the visibility of these objects will be related to aerosols and water vapor. These images, too, can be analyzed with *ImageJ*, by measuring grayscale brightness along a line drawn vertically upward from just below the horizon, or by calculating the average brightness of a selected uniform area within a distant object.

As with the aureole photos, you should use the same camera with the same manually set exposure and f-stop to allow comparisons under different conditions. It might be interesting to compare photos taken with fixed exposure and f-stop with photos of the same scene at the same time taken with automatic camera settings. The automatic settings will probably produce images that look "better" than images taken with manual settings because your camera will try to produce an image that is "properly" exposed, as defined by its light-processing algorithms.

Digital photography has progressed very rapidly in the first few years of the 21st century. The same inexpensive cameras and software that make digital photography so much fun can also be used to do some very interesting science. Professional atmospheric scientists have perhaps not yet thought enough about the opportunities this technology provides. As a result, this is an easily accessible area of investigation that can be very rewarding for students, teachers, and other non-specialists. Even individual photos of atmospheric conditions can be interesting, but as with all kinds of atmospheric measurements, consistent long-term records will be much more valuable.

6. Concluding Remarks

This book has provided an introduction to the sun, how Earth's atmosphere affects the transmission of sunlight to the surface, and how to design relatively simple instruments to measure solar radiation and, indirectly, some important atmospheric and surface properties. The intended audience is educators, students, and other non-specialists—in fact, anyone who is curious about how the sun/Earth/atmosphere system works.

The instruments described here are, at the very least, excellent instructional tools that will provide their users with a sense of how the measurement process works. This process is the same regardless of whether an instrument costs $20 or $20,000, so this is an authentic science experience. Although there is certainly a place for pre-packaged, "user-friendly" measurement tools and projects, a completely different experience and perspective on science is obtained when you construct your own instruments and design your own experiments. You will also gain a better appreciation of the essentially analog world of scientific measurement, a perspective that is especially important in the overwhelmingly digital environment to which all of us are continuously exposed.

It is important not to underestimate the difficulty of collecting reliable and accurate data in any area of science. It requires commitment, patience, and great care. In formal educational settings, the growing need to meet very specific teaching and learning standards in order to perform well on standardized tests means that finding time for innovative hands-on, inquiry-based activities presents a serious challenge for even the most determined teachers. In informal settings, it is still difficult to carve time from daily life to make the long-term commitments required to collect scientifically useful data.

These challenges aside, it is also important not to underestimate the value of data collected with the relatively inexpensive instruments described here. If they are carefully constructed, properly calibrated, and used under appropriate conditions, these instruments can produce (and have produced) very reliable, scientifically useful data. Ongoing testing with these instruments show consistently good agreement with their more expensive counterparts. In particular, the performance of the visible-light sun photometer described in Chapter 5, which has been in use longer than the other instruments described here, has been confirmed through extensive comparisons with other sun photometers, such as the CIMEL sun photo-meters that comprise the Aerosol Robotic Network (AERONET) [Holben *et al.*, 1998; Brooks *et al.*, 2003b]. The pyranometer discussed in Chapter 4 has also been subjected to extensive testing at several sites around the United States.

Because of the importance scientists attach to matters of instrument calibration and data collection procedures, it is a good idea to develop partnerships with scientists who can help to ensure the quality of data collected with these instruments. Being careful about data quality is not just a "scientific" issue. Measurements that cannot be compared with accepted standards will provide more confusion and disillusionment than they will enlightenment in an educational setting.

Most scientists are eager to help non-specialists understand their work and appreciate these kinds of collaborations, even though they may have little experience working with students and teachers. Most science educators, regardless of their teaching skills, lack research interests and experience. As a result, they are generally poorly prepared to show students how to develop their own research agendas and interpret what they find. Hence, partnerships among scientists, educators, and students are extremely valuable for bridging the gaps among these groups.

It is rare to find adults, including educators, who have the basic electronic construction skills needed to build the instruments described in this book. Nonetheless, this is a worthwhile and empowering experience for students *and* educators. There are a number of books that present the basics of electronic construction. (See, for example, Mims [2003b].) With adequate supervision, this kind of work is certainly appropriate for secondary and some middle school students.

It is usually inefficient and relatively expensive to purchase parts for making just one or two instruments, because electronics suppliers invariably favor purchases in quantity. In any event, it is a better idea to get enough parts for several instruments, especially in situations where inexperience will invariably lead to mistakes. In some cases, detailed instructions and/or complete kits of parts may be available through the author of this book, who should be easy to locate online. In many circumstances, this a good option whenever it is available.

Once instruments have been built, tested, and calibrated, what then? There is no reason why "amateurs" cannot establish their own solar/atmosphere observatories. Of course, not all sites are suitable for all the measurements described in this book. For example, it is not possible to measure full-sky solar radiation in an urban environment surrounded by tall buildings. However, this does not mean that pyranometer data collected in such an environment are without value. Collected consistently, such data will demonstrate clearly and dramatically the effects of shadowing and seasonal changes on the availability of sunlight in an urban environment.

On the other hand, direct sun measurements, including the aureole photography discussed at the end of Chapter 5, can be taken anywhere an unobstructed view of the sun is available.

In the context of this book, a "solar/atmosphere observatory" is as much a mindset as it is a physical place. What is important is an understanding of the opportunities and challenges associated with a particular site, a consistent approach to data collection and recordkeeping, and dedication to maintaining data continuity and quality. As long as these requirements are met, anyone can make valuable contributions to understanding our home planet.

Appendices

Appendix 1: List of Symbols

Because some symbols have different meanings depending on context, they are listed by chapter and appendix number, starting with Chapter 2.

Chapter 2

A	albedo (broadband reflectivity)	dimensionless
AU	astronomical unit	$\approx 1.5 \times 10^{11}$ m
e	eccentricity of an orbit, ≥ 0	dimensionless
km	unit of distance	kilometers
K	unit of absolute temperature	kelvins
m	unit of distance	meters
P	power	watts
r	Earth's mean equatorial radius	≈ 6378 km
R	average Earth-sun distance	m or AU
S	solar irradiance	W/m^2
S_o	solar irradiance at the average Earth/sun distance	W/m^2
T	absolute temperature	kelvins ($0°C = 273.15$ K)
W	unit of power	watts
x	"greenhouse factor"	dimensionless, $0 \leq x < 1$
σ	Stefan-Boltzmann constant	5.67×10^{-8} W/(m²K⁴)

Chapter 3

A, B, C, D	coefficients used to describe Rayleigh scattering	
g	referring to gasses (as subscript)	
I	intensity of radiation	arbitrary units
m_{air}	relative air mass	dimensionless, ≥ 1
p	atmospheric pressure	millibars
R	referring to Rayleigh scattering (as subscript)	
T	spectral transmission of sunlight	percent
z	solar zenith angle	degrees or radians

| α | optical thickness | dimensionless, >0 |
| λ | wavelength | microns (μm), in eqn. (3.6), usually in nm |

Chapter 4

A	unit of current	amperes
C	calibration constant	$(W/m^2)/V$
I	unit of current	amperes
mA	current	milliamperes
P	power	watts
R	resistance	ohms
V	unit of voltage	volts
W	unit of power	watts
Ω	unit of resistance	ohms

Chapter 5

a	referring to aerosols (as subscript)	
a, b	coefficients for T_{WV} (5.12 and 5.13)	
A, B, C	coefficients for water vapor instrument calibrations	
C_{Smith}	coefficient for calculating PW	dimensionless
e_s	saturated vapor pressure of atmosphere	millibars
h	site elevation	km
I	intensity of radiation	arbitrary units
m_{air}	relative air mass	dimensionless, ≥ 1
mb	unit of pressure	millibars
p	atmospheric pressure	millibars
PW	precipitable water vapor	cm of water (cm H_2O)
R	Earth-sun distance	astronomical units (AU)
R	referring to Rayleigh scattering (as subscript)	
RH	relative humidity	percent
T	air temperature	degrees Celsius
T_{WV}	water vapor transmission factor	dimensionless
T_d	dewpoint temperature	degrees Celsius
V	instrument output voltage	volts
V_o	sun photometer calibration constant	volts
z	solar zenith angle	degrees or radians
α	optical thickness	dimensionless, >0
β	exponent in definition of T_{WV}	dimensionless

Appendix 2

c	speed of light	2.9979×10^8 m/s
E	radiated power per unit area	W/m^2
h	Planck's constant	6.6261×10^{-34} Joule-s
k	Boltzmann's constant	1.38065×10^{-23} Joule/K
r	solar radius	6.96×10^8 m
R	average Earth-sun distance	1 AU, $1.5 \ 10^{11}$ m
T	absolute temperature	kelvins
λ	wavelength	m
Δλ	wavelength interval	m

Appendix 3

A	albedo (broadband reflectivitiy)	dimensionless
S$_o$	solar irradiance at the average Earth/sun distance	W/m^2
T	absolute temperature	kelvins
x	"greenhouse factor"	dimensionless, $0 \le x < 1$
σ	Stefan-Boltzmann constant	5.67×10^{-8} W/(m^{2}K^4)

Appendix 4

A, B	constants in equation for finding Julian Date	
AZ	azimuth angle of sun	degrees
JD	Julian Date	days
C	sun's equation of center	degrees
d	day of month	decimal number, 1–31
D	declination of sun, Earth latitude/longitude coordinates	degrees
e	eccentricy of Earth's orbit	dimensionless
E	equation of time	radians or degrees
f	true anomaly of the sun	degrees
HA	hour angle of sun	degrees
L$_o$	geometric mean longitude of the sun	degrees
L$_{true}$	true longitude of the sun	degrees
m	month of year	integer, 1–12
m$_{air}$	relative air mass	dimensionless
M	mean anomaly of the sun	degrees
OB	obliquity (angle of equatorial plane to ecliptic)	degrees

R   Earth/sun radius   astronomical units
RA   right ascension of sun, Earth latitude/
           longitude coordinates   degrees
ST   sidereal time   degrees
T   Julian centuries since January, 2000
y   year   four-digit integer
z   solar zenith angle   degrees or radians
ε   solar elevation angle   degrees or radians

Appendix 5

A   aerosol transmission factor   dimensionless
AU   astronomical unit $\approx 1.5 \times 10^{11}$ m
R   Earth-sun distance   AU
S   insolation W/m^2
S_o solar constant W/m^2
T_a, T_g, T_r, T_W
        transmission coefficents for aerosols,
        gasses, molecular scattering,
        and water vapor   dimensionless, $>0, \leq 1$

Appendix 7

pf   capacitance   picofarads
µf   capacitance   microfarads

Appendix 8

α   aerosol optical thickness   dimensionless, >0
β   Ångstrom turbidity coefficient   dimensionless, >0
λ   wavelength   microns, µm
τ   Ångstrom exponent   dimensionless

Appendix 2: Planck's Equation for Blackbody Radiation

As shown previously in Figure 2.1, the sun behaves approximately like a blackbody—a perfect absorber and radiator—at a temperature of about 5,800 K. The blackbody curve shown along with the extraterrestrial insolation in Figure 2.1 is calculated from Planck's law [Planck, 1901]:

$$E = \Delta\lambda \cdot (2\pi hc^2\lambda^{-5})/[\exp(hc/\lambda kT) - 1] \tag{A2.1}$$

where E is radiated power per unit area in the wavelength interval $\Delta\lambda$; λ is wavelength in meters; c is the speed of light, 2.9979×10^8 m/s; h is Planck's constant, 6.6261×10^{-34} Joule-s; k is Boltzmann's constant, 1.38065×10^{-23} Joule/K; T is temperature in kelvins.

The total power radiated in the interval $\Delta\lambda$ is $4\pi r^2 E$, where the sun's radius r is about 6.96×10^8 m. At the average Earth-sun distance R, the extraterrestrial radiation in the interval $\Delta\lambda$ is $E \cdot (r^2/R^2)$ W/m^2, where R is about 1.5×10^{11} m.

Planck derived this famous equation in an attempt to reconcile the behavior of blackbody radiation as described by the 19th-century Rayleigh-Jeans and Wien's laws. The Rayleigh-Jeans law explained blackbody radiation at long wavelengths, but not at short wavelengths. Wien's law worked for shorter wavelengths, but not for longer wavelengths. The Rayleigh-Jeans law had a firm foundation in classical electromagnetic theory as it was understood at the time, so its breakdown at short wavelengths (the "ultraviolet catastrophe") was profoundly disturbing to physicists.

When Planck searched for a mathematical description that would work for all wavelengths, he found that the available data on blackbody radiation could be explained by assuming that radiation is emitted only in discrete packets with an energy proportional to the inverse of wavelength: energy $= hc/\lambda$. Planck considered this assumption to be only a "fudge factor" that provided an empirical explanation of blackbody radiation. However, other physicists soon realized that this assumption must, in fact, have a physical basis that required a fundamentally new theory of electromagnetic radiation.

In 1905, Albert Einstein published a Nobel-prize-winning paper [Einstein, 1905] showing that the well-known photoelectric effect, in which light striking certain surfaces causes a small current to flow, could not be explained by classical theories of electromagnetic radiation, but *can* be explained by assuming that light energy is quantized—transmitted only in discrete units by what are now called photons.

However, classical theories, which attribute continuous wavelike behavior to light and other forms of electromagnetic radiation, still provide perfectly workable explanations for phenomena such as the interference patterns caused by light passing through small slits. The realization that electromagnetic radiation must have both wavelike and particle-like properties, no matter how counterintuitive such a conclusion seemed, revolutionized physics and led to the development of what is now known as quantum mechanics.

Appendix 3: Design Your Own Planet

The basic equation describing the radiative balance for the Earth/ atmosphere system as viewed from space has been given in Chapter 2:

$$4\Phi T^4 = S_0(1 - A) \qquad\qquad\qquad (A3.1)$$

Now consider a very simple model of the Earth/atmosphere system that reduces the atmosphere to a single homogeneous layer. This atmosphere absorbs some of the energy emitted by Earth's surface. Some of this energy is re-radiated to space and some is radiated back to Earth's surface, thereby changing the radiative balance:

$$4\Phi T^4(1 - x) = S_0(1 - A) \qquad\qquad\qquad (A3.2)$$

where x, with a value between 0 and 1, is a measure of the greenhouse effect of the atmosphere. T (in kelvins) is now not the apparent temperature of the Earth/atmosphere system as viewed from space but, in a very simplified view, the temperature of a single-layer atmosphere. The larger the value of x (that is, the more of Earth's thermal radiation is reradiated back to Earth's surface rather than outward to space) the higher T must be to maintain the single-layer atmosphere in radiative equilibrium.

For $x = 0$, there is no absorption and no greenhouse effect; that is, the atmosphere is transparent to both incoming solar radiation and outgoing thermal radiation, as though it weren't even there. A value of $x = 0.4$ produces the actual equilibrium surface temperature for the Earth/ atmosphere system of about 16°C. Values closer to 1 lead to a "runaway greenhouse effect," such as exists on Venus, resulting in very high surface temperatures.

In order to design your own planet, you need to be able to vary the planet/sun distance, the average planetary reflectivity (albedo) A, and the greenhouse parameter x. Table A3.1 shows a model implemented in a spreadsheet. It allows the user to specify three surfaces—land, water, snow/ice—plus clouds, the planetary coverage for each type, and an albedo for each type. Cloud cover is assumed to reduce each of the three surface types by the same amount. For example, if 30% of the planet is land and the average cloud cover is 50%, this reduces the land cover to 15%.

The fractions of land, water, and snow/ice must sum to 1. Based on these assumptions, the planetary albedo is:

$$A_{planet} = land \cdot (1 - cloud) \cdot A_{land} + water \cdot (1 - cloud) \cdot A_{water}$$
$$+ snow\_ice \cdot (1 - cloud) \cdot A_{snow\_ice} + cloud \cdot A_{cloud} \qquad (A3.3)$$

The spreadsheet model allows the user to vary separately the planet/sun distance, the fraction of each surface type, the fraction of cloud cover, the greenhouse factor x, and the solar constant S. Figures A3.1–A3.3 show some typical results. The albedo values used are: land = 0.10, water = 0.02, snow = 0.80, cloud = 0.75. These values yield a planetary albedo of about 0.30, Earth's albedo as viewed from space, for a planet at the same distance from the sun as Earth.

Table A3.1. Spreadsheet model for "designing" a planet.

T as a function of greenhouse effect parameter (x)								
		Surface types						
Planet/sun distance	Land	Water	Snow/ Ice	Cloud cover	x	S	Albedo	T
1.50E + 08	0.25	0.65	0.1	0.30	0.00	1379.3	0.3076	-18.3
1.50E + 08	0.25	0.65	0.1	0.30	0.05	1379.3	0.3076	-15.0
1.50E + 08	0.25	0.65	0.1	0.30	0.10	1379.3	0.3076	-11.5
1.50E + 08	0.25	0.65	0.1	0.30	0.15	1379.3	0.3076	-7.7
1.50E + 08	0.25	0.65	0.1	0.30	0.20	1379.3	0.3076	-3.6
1.50E + 08	0.25	0.65	0.1	0.30	0.25	1379.3	0.3076	0.7
1.50E + 08	0.25	0.65	0.1	0.30	0.30	1379.3	0.3076	5.5
1.50E + 08	0.25	0.65	0.1	0.30	0.35	1379.3	0.3076	10.7
1.50E + 08	**0.25**	**0.65**	**0.1**	**0.30**	**0.40**	**1379.3**	**0.3076**	**16.4**
1.50E + 08	0.25	0.65	0.1	0.30	0.45	1379.3	0.3076	22.8
1.50E + 08	0.25	0.65	0.1	0.30	0.50	1379.3	0.3076	29.9
1.50E + 08	0.25	0.65	0.1	0.30	0.55	1379.3	0.3076	38.0
1.50E + 08	0.25	0.65	0.1	0.30	0.60	1379.3	0.3076	47.3
1.50E + 08	0.25	0.65	0.1	0.30	0.65	1379.3	0.3076	58.2
1.50E + 08	0.25	0.65	0.1	0.30	0.70	1379.3	0.3076	47.3
1.50E + 08	0.25	0.65	0.1	0.30	0.75	1379.3	0.3076	87.3
1.50E + 08	0.25	0.65	0.1	0.30	0.80	1379.3	0.3076	107.9
1.50E + 08	0.25	0.65	0.1	0.30	0.85	1379.3	0.3076	136.3
1.50E + 08	0.25	0.65	0.1	0.30	0.90	1379.3	0.3076	180.0
1.50E + 08	0.25	0.65	0.1	0.30	0.95	1379.3	0.3076	265.7

Table A3.1. (continued)

T as a function of planet/sun distance

Planet/sun distance	Land	Water	Snow/ Ice	Cloud cover	x	S	Albedo	T
			Surface types					
5.00E + 07	0.25	0.65	0.1	0.30	0.45	12414.1	0.3076	239.3
7.50E + 07	0.25	0.65	0.1	0.30	0.45	5517.4	0.3076	145.3
1.00E + 08	0.25	0.65	0.1	0.30	0.45	3103.5	0.3076	89.3
1.25E + 08	0.25	0.65	0.1	0.30	0.45	1986.3	0.3076	51.0
1.50E + 08	0.25	0.65	0.1	0.30	0.45	1379.3	0.3076	22.8
1.75E + 08	0.25	0.65	0.1	0.30	0.45	1013.4	0.3076	0.9
2.00E + 08	0.25	0.65	0.1	0.30	0.45	775.9	0.3076	-16.8
2.25E + 08	0.25	0.65	0.1	0.30	0.45	613.0	0.3076	-31.5
2.50E + 08	0.25	0.65	0.1	0.30	0.45	496.6	0.3076	-43.9
2.75E + 08	0.25	0.65	0.1	0.30	0.45	410.4	0.3076	-54.5
3.00E + 08	0.25	0.65	0.1	0.30	0.45	344.8	0.3076	-63.8
3.25E + 08	0.25	0.65	0.1	0.30	0.45	293.8	0.3076	-72.0
3.50E + 08	0.25	0.65	0.1	0.30	0.45	253.3	0.3076	-79.4
3.75E + 08	0.25	0.65	0.1	0.30	0.45	220.7	0.3076	-85.9
4.00E + 08	0.25	0.65	0.1	0.30	0.45	194.0	0.3076	-91.9
4.25E + 08	0.25	0.65	0.1	0.30	0.45	171.8	0.3076	-97.3
4.50E + 08	0.25	0.65	0.1	0.30	0.45	153.3	0.3076	-102.2
4.75E + 08	0.25	0.65	0.1	0.30	0.45	137.6	0.3076	-106.8
5.00E + 08	0.25	0.65	0.1	0.30	0.45	124.1	0.3076	-111.0
5.25E + 08	0.25	0.65	0.1	0.30	0.45	112.6	0.3076	-114.9
5.50E + 08	0.25	0.65	0.1	0.30	0.45	102.6	0.3076	-118.5
5.75E + 08	0.25	0.65	0.1	0.30	0.45	93.9	0.3076	-121.9
6.00E + 08	0.25	0.65	0.1	0.30	0.45	86.2	0.3076	-125.1
6.25E + 08	0.25	0.65	0.1	0.30	0.45	79.5	0.3076	-128.1

Table A3.1. (continued)

T as a function of cloud cover

Surface types

Planet/sun distance	Land	Water	Snow/ Ice	Cloud cover	x	S	Albedo	T
1.50E + 08	0.25	0.65	0.1	0.00	0.40	1379.3	0.1180	34.5
1.50E + 08	0.25	0.65	0.1	0.05	0.40	1379.3	0.1496	31.7
1.50E + 08	0.25	0.65	0.1	0.10	0.40	1379.3	0.1812	28.8
1.50E + 08	0.25	0.65	0.1	0.15	0.40	1379.3	0.2128	25.9
1.50E + 08	0.25	0.65	0.1	0.20	0.40	1379.3	0.2444	22.8
1.50E + 08	0.25	0.65	0.1	0.25	0.40	1379.3	0.2760	19.7
1.50E + 08	0.25	0.65	0.1	0.30	0.40	1379.3	0.3076	16.4
1.50E + 08	0.25	0.65	0.1	0.35	0.40	1379.3	0.3392	13.1
1.50E + 08	0.25	0.65	0.1	0.40	0.40	1379.3	0.3708	9.6
1.50E + 08	0.25	0.65	0.1	0.45	0.40	1379.3	0.4024	6.0
1.50E + 08	0.25	0.65	0.1	0.50	0.40	1379.3	0.4340	2.2
1.50E + 08	0.25	0.65	0.1	0.55	0.40	1379.3	0.4656	-1.7
1.50E + 08	0.25	0.65	0.1	0.70	0.40	1379.3	0.4972	-5.8
1.50E + 08	0.25	0.65	0.1	0.65	0.40	1379.3	0.5288	-10.1
1.50E + 08	0.25	0.65	0.1	0.70	0.40	1379.3	0.5604	-14.6
1.50E + 08	0.25	0.65	0.1	0.75	0.40	1379.3	0.5920	-19.4
1.50E + 08	0.25	0.65	0.1	0.80	0.40	1379.3	0.6236	-24.5
1.50E + 08	0.25	0.65	0.1	0.85	0.40	1379.3	0.6552	-29.9
1.50E + 08	0.25	0.65	0.1	0.90	0.40	1379.3	0.6868	-35.6
1.50E + 08	0.25	0.65	0.1	0.95	0.40	1379.3	0.7184	-41.9

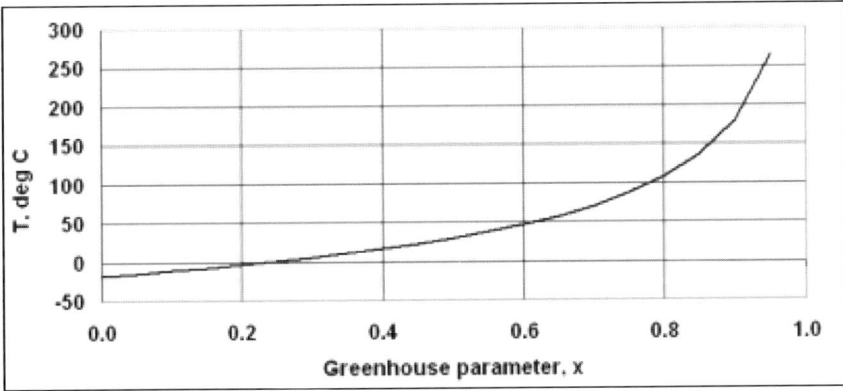

A3.1a. Surface temperature as a function of the greenhouse parameter.

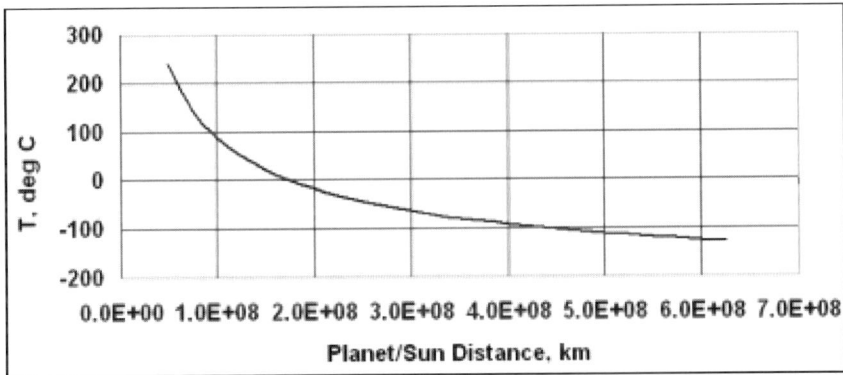

A3.1b. Surface temperature as a function of planet/sun distance.

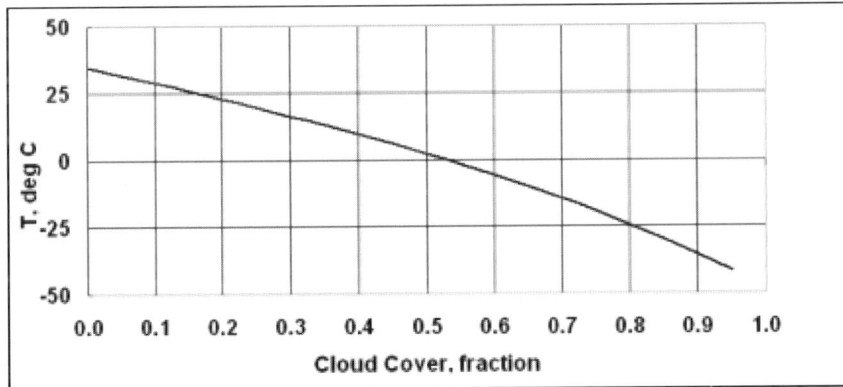

A3.1c. Surface temperature as a function of cloud cover.

Figure A3.1. "Designing" a planet with variable greenhouse effects, planet/sun distance, and cloud cover.

Appendix 4: Where Is the Sun?

Many measurements of solar transmission through the atmosphere, specifically all sun photometer measurements, require that the position of the sun above the horizon (its elevation angle) or, equivalently, its angular position from the zenith (90° minus the elevation angle), be accurately known. This may be required even for full-sky measurements, as their interpretation may depend on solar position.

It is possible to measure the solar elevation or zenith angle directly by measuring the length of a shadow cast on a horizontal surface by a vertical rod of known length. This is a useful exercise even for young students. It can be carried out at several different times during a year and can be used to find the maximum solar elevation and time of solar noon as the seasons change. However, it is certainly less time-consuming to computerize calculations of solar position based on the date, time, and longitude-latitude coordinates at which a measurement is taken. This reduces the work required of an observer and can be done at any time before or after a measurement.

The appropriate application of astronomical equations makes this a very accurate calculation—more accurate than can be obtained by a casual observation. The calculations are not simple, but at least the computer coding needs to be done only once. The equations presented below are from a well-known book on astronomical calculations by Belgian author Jean Meeus [1991]. For more detailed explanations of the equations and variables used below, consult this or a similar reference.

The first requirement for locating the sun is to convert a calendar date to its equivalent Julian day (JD).[1] The Julian day provides a unique value for every day since several thousand years B.C., and is required for the calculations that follow. Given a date expressed as a month, m, day, d, and 4-digit year, y,

If $m \leq 2$, subtract 1 from the year and add 12 to the month

Define $A = <y/100>$, $B = 2 - A + <A/4>$

$$JD = <365.25 \cdot (y+4716)> + <30.6001 \cdot (m+1)> + d + B - 1524.5 \qquad (A4.1)$$

---

[1] Sometimes, the "Julian day" or "Julian date" is interpreted as the day number during the calendar year, with values from 1 to 365 or 366, but this is an entirely different quantity than the Julian day as defined here.

Expressions written as <...> are interpreted as "the truncated (*not* rounded) integer value of..." Thus, <8/3> = 2. Julian days start at noon, Greenwich Mean Time (Universal Time).[2] Here is an example:

m = 11, d = 30, y = 2003, UT = 00:00:00 (start of calendar day)

A = <2003/100> = 20 B = 2 − 20 + 5 = −13

JD = <365.25·6719> + <30.6001·12> + 30 − 13 − 1524.5
    = 2454114 + 367 + 30 − 13 − 1524.5 = 2452973.5

The calendar day, d, can have a fractional part. If d = 30.75, the fractional part corresponds to (0.75)(24) = 18 hours, or 6 p.m. on November 30. The Julian day for this time is 2452973.5 + 0.75 = 2452974.25.

The next step is to calculate the position of the sun as viewed from Earth. The equations presented here are sufficient to write computer code to do the calculations, but it is beyond the scope of this discussion to give a complete description of the terminology and details of the calculations. For more details, refer to Meeus [1991] or online sources dealing with astronomical calculations.

The calculations defined here imply an accuracy that may look like "overkill" for this problem. However, even simpler versions of these calculations require a computerized implementation in a program or spreadsheet. So, once the initial work of entering the equations has been done, there is no extra work associated with calculations done to this precision. Also, less precise representations of the equations can lead to significant errors when positions are projected far into the future or past.

The first set of calculations gives the Earth/sun distance (R) and the longitude of the sun in ecliptic coordinates. (The ecliptic plane is the plane in which Earth rotates around the sun.)

A Julian century is 36,525 days. So, the number of Julian centuries from January 2000 is

$$T = (JD − 2451545.0)/36525.0 \qquad\qquad (A4.2)$$

The next step is to find the sun's geometric mean longitude in ecliptic coordinates:

$$L_o = 280.46645 + 36000.76983T + 0.0003032T^2 \qquad (A4.3)$$

The mean anomaly of the sun is

$$M = 357.52910 + 35999.05030T - 0.0001559T^2 - 0.00000048T^3 \quad (A4.4)$$

The eccentricity of Earth's orbit is

$$e = 0.016708617 - 0.000042037T - 0.0000001236T^2 \qquad (A4.5)$$

The equation of center for the sun, relative to its mean anomaly, is

$$C = (1.914600 - 0.004817T - 0.000014T^2) \cdot \sin(M)$$
$$+ (0.019993 - 0.000101T) \cdot \sin(2M) + 0.000290 \cdot \sin(3M) \qquad (A4.6)$$

from which its true longitude in ecliptic coordinates is

$$L_{true} = (L_o + C) \bmod 360 \qquad (A4.7)^3$$

Its true anomaly is

$$f = M + C \qquad (A4.8)$$

from which the Earth-sun radius is

$$R = 1.000001018 \cdot (1 - e^2)/[1 + e \cdot \cos(f)] \qquad (A4.9)$$

These equations give angular values in degrees for L_o, M, C, L_{true}, and f. However, trigonometric functions in computer programming languages and spreadsheet functions typically expect angles to be expressed in radians, not degrees. So, when M and f are used as arguments in a trigonometric function, as they are in Equations (A4.6) and (A4.9), for example, they should almost certainly be converted to radians:

$$radians = (degrees)(\pi/180) \qquad (A4.10)$$

Next, calculate the mean sidereal time—the angular position (hour angle) of the point at which the ecliptic plane intersects the equator.

---

[3] y mod x is the remainder of dividing y by x, e.g., 11.5 mod 7 = 4.5.

$$ST = 280.46061837 + 360.98564736629 \, (JD - 2451545)$$
$$+ \, 0.000387933 \, T^2 - T^3/38710000$$
$$ST \rightarrow ST \bmod 360 \tag{A4.11}$$

where the "mod" operation returns the remainder from dividing "sidereal" by 360. For example, 733.37 mod 360 = 13.37.

Next, calculate the obliquity of the ecliptic, the angle between the equatorial and the ecliptic planes:

$$OB = 23 + 26/60 + 21.448/3600 - (46.8150/3600)T$$
$$- \, (0.00059/3600)T^2 + (0.001813/3600)T^3 \tag{A4.12}$$

Finally, convert the sun's position from ecliptic coordinates to right ascension, declination, hour angle, and azimuth in Earth equatorial (longitude and latitude) coordinates:

$$RA = \arctan[\tan(L_{true}) \cos(OB)] \tag{A4.13}$$

$$D = \arcsin[\sin(OB) \sin(L_{true})] \tag{A4.14}$$

$$HA = ST + Lon_{observer} - RA \tag{A4.15}$$

$$AZ = \arctan(y/x) \tag{A4.16}$$

where

$$y = \sin(HA)$$
$$x = \cos(HA) \sin(Lat_{observer}) - \tan(D) \cos(Lat_{observer})$$

The warning about trigonometric functions expecting angles expressed in radians still applies. Note that the azimuth needs to be expressed as a value between 0° and 360°, or −180° and 180°. If the tangent of an angle is expressed simply as the value y/x, either positive or negative, there is ambiguity about the quadrant. Some programming languages and spreadsheets include arctangent functions that require both the x and y coordinates to be specified so the function can consider the sign of each component to determine the proper quadrant for the result.[4] The solar elevation, ε, and zenith angle, z, are:

---

[4] In C, for example, the function atan(x) returns a value between ±π/2 radians. The atan2(y,x) function returns a value between ±π radians based on the individual signs of x and y.

$$\varepsilon \text{ (radians)} = \arcsin[\sin(\text{Lat}_{\text{observer}}) \sin(D)$$
$$+ \cos(\text{Lat}_{\text{observer}}) \cos(D) \cos(HA)] \qquad \text{(A4.17)}$$

$$z = 90° - \varepsilon \text{ (degrees)} \qquad \text{(A4.18)}$$

The relative air mass, m_{air}, which is always required to process sun photometer data, is defined as shown in Figure A4.1 by the solar elevation or zenith angle. It is approximately equal to $1/\cos(z)$. It has a value of 1 at $z = 0°$ and increases with increasing zenith angle. At a zenith angle of 90°, this expression is no longer mathematically

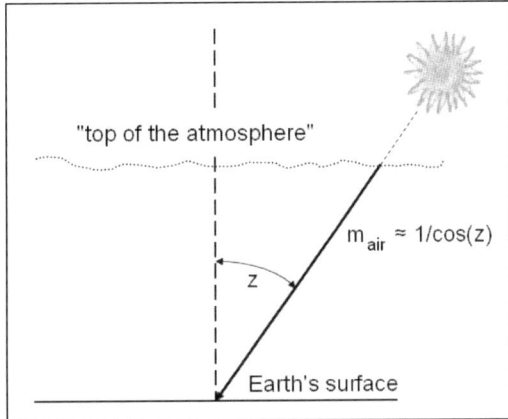

Figure A4.1. Relative air mass.

defined. Also, as the sun gets closer to the horizon, Earth's curvature and the refraction of the atmosphere require corrections to this simple formulation. For example, because of atmospheric refraction, the sun is still visible even when it is physically (geometrically) below the horizon. Here is a formulation by Young [1994], in terms of solar zenith angle, which takes into account curvature and refraction and is mathematically well-behaved at $z = 90°$:

$$m_{\text{air}} = [1.002432 \cos(z)^2 + 0.148386 \cos(z) + 0.0096467]/ [\cos(z)^2$$
$$+ .149864 \cos(z)^2 + 0.0102963 \cos(z) + 0.000303978] \qquad \text{(A4.19)}$$

The large number of significant digits represented in the equation guarantees the "good behavior" of the function when the sun is near the horizon. However, as is often the case with this kind of empirical representation, this does not imply that the calculation of m_{air} is actually that precise.

Figure A4.2 shows solar elevation and azimuth angle, plus relative air mass, as a function of Universal Time for a location at 75° W longitude and 40° N latitude (Philadelphia, PA, USA), on 21 June 2003, near the summer solstice. Azimuth is measured clockwise from north. In the summer, the sun at 40° N latitude rises north of due east. The sun crosses the local meridian at local solar noon (by definition) at an elevation angle of about 73.5°. In general, local solar noon is not the same as clock noon. The sun is still south of the site, because the solar declination is never more than about 23.5° north of the equator. (90° − (40° − 23.5°)) = 73.5°. Based on its definition, azimuth changes sign at noon and becomes −180° as it crosses the observer's meridian. In the evening, the sun sets north of due west. At local solar noon, the relative air mass has a minimum value of 1/sin (73.5°) = 1.043; it never reaches a value of 1 because the sun is never directly overhead at 40° N latitude.

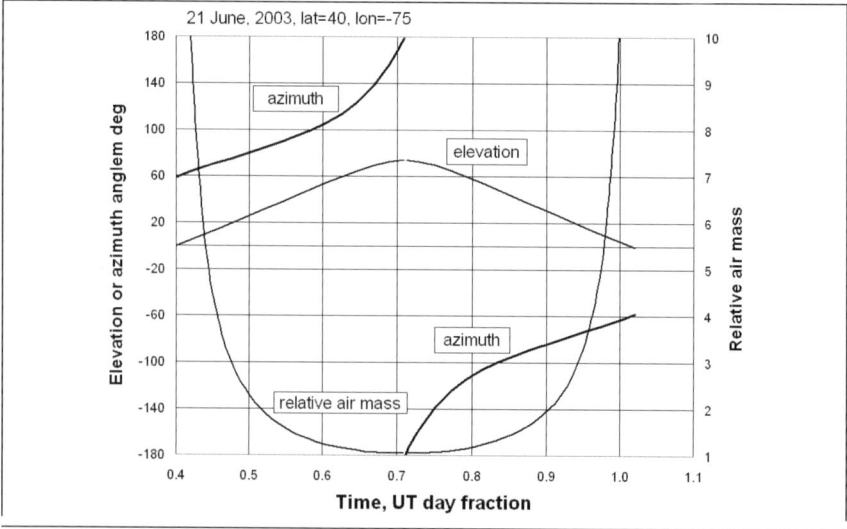

Figure A4.2. Solar azimuth and azimuth angles, and relative air mass near Philadelphia, Pennsylvania, USA, 21 June 2003.

It should be clear that local solar noon cannot be the same as local clock noon everywhere in the same time zone, because Earth rotates under the solar meridian at different times, depending on longitude. However, local solar noon is, in general, never the same as clock noon, regardless of longitude. The reason is that Earth's orbit around the sun is slightly elliptical, causing the apparent motion of the sun around Earth to appear faster when Earth is closer to the sun and slower when Earth is farther from the sun.

Clock time is based on a fictitious "mean" sun—a mathematical construction rather than a real object—which "rotates" around Earth. This mean sun has the same apparent orbital period as the real sun (one year), but lags or leads the actual sun during the year. The so-called equation of time describes the relationship between the real and the mean sun. Meeus [1991] gives this calculation for the equation of time E, in radians:

$$y = \tan^2(OB/2)$$
$$E = y\cdot\sin(2L_o) - 2e\cdot\sin(M) + 4ey\cdot\sin(M)\cos(2L_o)$$
$$- 0.5y^2\sin(4L_o) - 1.25e^2\sin(2M) \qquad\qquad (A4.20)$$

It may seem strange to express "time" as an angle. However, this is the conventional terminology. Time and angle are related due to the fact that Earth rotates 360° in 24 h of mean solar time. So, 15° equals one clock hour. To express the equation of time in minutes, convert E to degrees (multiply by 180/π), divide by 15 to get the fraction of an hour, and multiply by 60. The equation of time varies very slightly from year to year, but the equation of time for 2003 shown in Figure A4.3, in which E is plotted in units of clock minutes, adequately illustrates the annual cycle for the early 21st century.

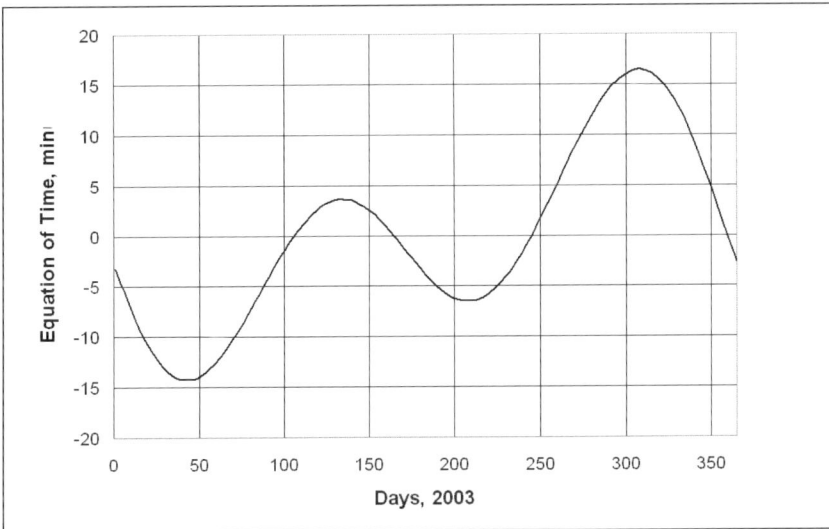

Figure A4.3. The equation of time.

To find local solar noon relative to clock noon at a particular longitude, first subtract the quantity (longitude mod 15) hours, where longitude is negative to the west. Then subtract the equation of time value E.

Here is an example: When is local solar noon (standard time) at 80° W longitude on November 01, 2007?

$E = 0.071671$ radians $= (0.071671)(180/\pi)(60/15) = 16.426$ min
$(-80 \bmod 75) = -5$
local noon $= 12$ h $- (-5/15)(60) - 16.426$ min
               $= 12$ h $+ (20 - 16.426)$ min
               $= 12$ h $+ 3.574$ min $= 12{:}03{:}34$

Depending on where you are, you may need to apply daylight saving or other local time corrections. For example, during the months when daylight saving time is in effect on the east coast of the United States, local solar noon occurs around one o'clock in the afternoon rather than around noon.

Appendix 5: A Simple Model of Sunlight Transmission Through the Atmosphere

As noted in Chapter 4, there is no easily accessible absolute calibration source for pyranometers. Hence, the only alternatives are to compare pyranometer output against a reliable reference or to use a model to predict insolation. The simple model presented here provides such a model.

It is important to understand that this is only an *approximate* model that provides some insight into the basic physical processes that control the amount of sunlight reaching Earth's surface under "clear sky" (no cloud) conditions. Nonetheless, under appropriate conditions of clear and relatively dry and unpolluted skies, it is a useful tool for calibrating a pyranometer if no other option is available.

Insolation, the total solar irradiance on a horizontal surface at Earth's surface, is controlled primarily by the seasons and the weather. Maximum daily total insolation is greatest during the summer when the sun is highest in the sky. Even under clear sky conditions, the atmosphere reduces the amount of sunlight that reaches Earth's surface. Aerosols (small particles suspended in the atmosphere) scatter sunlight, some of which is returned to space. Gasses (including water vapor) and aerosols also absorb sunlight, and some of this energy is re-radiated back to space. Of course, clouds have a major effect on insolation because they reflect a great deal of sunlight back to space. Their effects on insolation at a particular place and time are essentially impossible to predict accurately, which is why this model applies only to cloud-free skies.

Although the conditions that determine insolation are difficult to model with high accuracy even under clear skies, it is possible to write a simple equation that accounts at least conceptually for the factors that determine clear sky insolation S:

$$S = S_o \cos(z) T_r T_g T_w T_a / R^2 \qquad (A5.1)$$

where S_o is the solar constant a 1 AU, z is the solar zenith angle, and R is the Earth-sun distance in Astronomical Units (average distance is 1 AU).

The transmission factors, T, are dimensionless values between 0 and 1 which account for reductions in transmission of sunlight through the atmosphere to Earth's surface. Duchon and O'Malley [1999] have given these factors for molecular scattering (also called "Rayleigh scattering," hence the "r" subscript), gas absorption and scattering, water vapor absorption, and aerosol absorption and scattering as:

$$T_r T_g = 1.021 - 0.084 \, [m_{air} \, (949 \cdot p \times 10^{-6} + 0.051)]^{1/2} \qquad (A5.2)$$

$$T_w = 1.0 - 0.077 \, (PW \cdot m_{air})^{0.3} \qquad (A5.3)$$

$$T_a = (A)^{m_{air}} \qquad (A5.4)$$

where p is barometric pressure in millibars, PW is total column precipitable water vapor in units of centimeters of H_2O, A is an aerosol transmission factor, and m_{air} is the relative air mass. The relative air mass is 1 when the sun is directly overhead and varies approximately as $1/\cos(z)$. (See Equation (A4.19) for a more accurate calculation.) The model assumes typical values for a so-called "standard atmosphere" that scientists use for modeling the behavior of the atmosphere. A water vapor content 1.42 cm of water is the value assumed for this standard atmosphere. In very dry or high-elevation locations, PW can be as low as a few millimeters. In very wet locations, it can be as high as 6 cm. In typical temperate climates, a value of 3 or 4 might be more appropriate. The solar constant in Equation (A5.1), with an average value of about 1,370 W/m^2, is adjusted as shown by dividing it by the square of the actual Earth-sun distance in astronomical units (so S will be larger than S_o when the sun is closer to Earth).

At any elevation above sea level, you need to use "station pressure"—the actual barometric pressure. With only a few exceptions for research sites, weather reports *always* give pressure converted to sea level pressure. If you are at a higher elevation, you need to convert this value to your elevation. An approximate conversion is:

station pressure = (sea level pressure) − (elevation in meters)/9.2   (A5.5)

That is, pressure decreases very roughly 1 mb for each 10 m increase in elevation. Alternatively, use Equation (5.4) from Chapter 5.

The aerosol transmission factor A in Equation (A5.4) is assigned a value of 0.935 by default, although this is probably too large for many actual atmospheres. You can change any of the input values given here if you like, but you may get very strange results if you change them arbitrarily!

Here is a reasonable sample calculation of sunlight reaching Earth's surface, assuming a somewhat dirtier and wetter summertime atmosphere:

Inputs:

S_o = 1370.0, A = 0.9, PW = 3.0, p = 1000, z = 30.0, R = 1.03

Calculated outputs:

$T_r T_g$ = 0.9307, T_w = 0.8882, T_a = 0.8854, S = 818.6

More sophisticated broadband insolation models are available [*e.g.*, Bird and Hulstrom, 1981], but, they too are clear sky models. The Bird and Hulstrom model allows direct input of aerosol optical thickness at 500 nm and 380 nm. If the sun photometer described in Chapter 5 is used to collect AOT data, those values, at wavelengths of 505 nm and 625 nm, can be converted to 500 nm and 380 nm values through the Ångstrom exponent, the calculation of which is described in Appendix 8.

A wavelength-dependent model of solar radiation and its transmission through the atmosphere (a "radiative transfer model") is critical for designing and understanding spectrally selective sunlight detectors. A "Simple Model of the Atmospheric Radiative Transfer of Sunshine (SMARTS2)" [Gueymard, 1995], is a relatively easy to use computer model, available online at no cost from the National Renewable Energy Laboratory [NREL, 2008]. This model was used to generate the solar irradiance data in Figure 2.1 (comparing total solar radiation at the top of the atmosphere to blackbody radiation) and Figure 5.7 (showing the effects of water vapor absorption in the atmosphere and the spectral response of near-IR detectors).

Appendix 6: Using a Data Logger

For some of the instruments described in this book, it makes the most sense to record their output continuously over an extended period of time. Historically, this kind of recording was done with a strip chart recorder that used a moving pen drawing on a piece of paper that moved slowly under the pen—a conceptually primitive analog recording device. More recently, almost all long-term data recording is done with digital data loggers that convert an analog voltage signal into a digital number that can be stored in electronic memory. Considering the reminders in this book that the world is an analog place, this may seem like a remarkable conceptual about-face, but the fact remains that digital data loggers facilitate measurements that would otherwise be expensive and/or impractical.

Data loggers interface directly only with voltage signals, not current, as noted in the discussion of pyranometers in Chapter 4. No matter how transparent the recording process is made to appear, through the use of "smart" loggers that self-configure themselves to interpret outputs from attached sensors, this basic fact remains unchanged. This observation is relevant to the sensors discussed in this book, because they are devices that produce current when they are exposed to light.

There are two primary specifications for a data logger: its full-scale voltage range and its digital resolution. The voltage range can be for positive voltages only, or for negative and positive voltages. These ranges could be specified as, for example, 0–2.5 V for a logger that will record only positive voltages, or ± 10 V for a logger that will record both positive and negative voltages.

The digital resolution specification is important because it indicates whether a logger is suitable for a particular task. The resolution is given as the number of bits used to represent an analog signal value that is converted to a digital number. Typical resolutions are 8-bit, 10-bit, 12-bit, and 16-bit. Not surprisingly, prices increase with the digital resolution.

To understand what these designations mean, consider how numbers are stored digitally. Computers work basically with bits—on or off states in electronic memory. An 8-bit "word" uses 8 on or off bits. Such a word can store integer values between 0 and 255. Each bit in the 8-bit word represents a power of 2, stored from right to left:

2^7	2^6	2^5	2^4	2^3	2^2	2^1	2^0

The integer values between 0 and 255 are represented as the sum of powers of 2. Here are some examples:

$255 = 11111111$
$\qquad = 2^7 + 2^6 + 2^5 + 2^4 + 2^3 + 2^2 + 2^1 + 2^0 = 128+64+32+16+8+4+2+1$
$7 = 00000111 = 2^2 + 2^1 + 2^0 = 4+2+1$
$119 = 01110111 = 2^6 + 2^5 + 2^4 + 2^2 + 2^1 + 2^0 = 64+32+16+4+2+1$

Although this may not seem to be a very efficient way to store numbers, computers are very good at handling operations based on binary arithmetic.

Consider a 0–2.5 V 8-bit logger. This means that an analog input between 0 and 2.5 V is stored as one of 256 inter values between 0 and $2^8 - 1 = 255$. So, the resolution of this logger is $2.5/255 = 0.0098 \approx 10$ mV. Is this good enough? Take the example of the solar-cell-based pyranometer discussed in Chapter 4, which has an output in full sunlight of about 200 mV. A 10-mV resolution means that the conditions from dark to full sun will be represented by only $200/10 = 20$ values: 0 mV, 10 mV, 20 mV, etc. Suppose that 200 mV represents 1000 W/m^2. This means that the resolution is 50 W/m^2, which is probably not sufficient for measuring insolation.

Now consider a 12-bit logger. The largest integer that can be represented is $2^{12} - 1 = 4095$. A 0–2.5 V, 12-bit logger has a full-scale resolution of $2.5/4095 = 0.00061$ V ≈ 0.6 mV. Again consider a pyranometer whose maximum output is about 200 mV. This range is represented by about $200/0.6 = 333$ values. If 200 mV corresponds to 1000/m^2, the resolution is $1000/333 = 3$ W/m^2.

For most purposes, 12-bit resolution will be adequate for this pyranometer, as the absolute radiometric uncertainty is no doubt larger than 3 W/m^2 under typical conditions. Although it is easy to say that more resolution is always better, the fact remains that at the time this book was written, the price difference between commercially available 12-bit and 14- or 16-bit loggers made it impractical to recommend higher resolution loggers for use with the inexpensive instruments described in this book. Rapidly advancing technologies may eliminate this argument in the near future.

It is possible to amplify small signals, as has been done for the LED detectors used in the sun photometers described in Chapter 4, to make better use of the range of a particular data logger. In that case, the very small current output from the LED detectors is converted to a voltage in the 1–2 V range by using a transimpedance amplifier, as described in Appendix 7 (the "active electronics" equivalent of forcing the solar cell to do work against a resistive load), so that its output in full sunlight is closer to the full-scale range of a data logger (e.g., 2.5 V rather than 200 mV). Its insolation resolution could then be $1000/4095 = 0.24$

W/m^2 rather than 3 W/m^2. However, amplifiers should be avoided, as they have been for the pyranometer discussed in Chapter 4, unless they are necessary. Amplifiers add cost and complications such as electronic noise and temperature sensitivity.

At the time this book was written, there were several manufacturers of 12-bit loggers. Some are intended for educational use and include "smart probes" that are very easy to use. However, they may be too expensive to dedicate to an instrument like a pyranometer, which provides the most interesting data when its output is recorded continuously over long periods of time.

There are two basic kinds of data logging systems—standalone loggers with internal memory, and analog-to-digital (A/D) converters that are connected directly to a computer, which serves as the storage medium.

For most the data logging needs of instruments described in this book, the author uses the standalone U12 series of 12-bit USB loggers from Onset Computer Corporation. They are relatively inexpensive (~$100) and extremely reliable if used under appropriate conditions.

The author has also used the 232SDA12 12-bit A/D converter from B&B Electronics (~$60). These are only A/D converters and must be connected to a computer for data storage, as they have no internal memory. Their advantage is that they support up to 11 channels of 0–5 V analog input, with a resolution of about 1.2 mV. They work very well with an ancient HP200LX handheld computer that runs MS-DOS, but this setup is antiquated, and these devices may be impractical for use with laptops that no longer include serial ports and run versions of Windows that no longer even support "DOS-like" operations. This device requires a separate power source. The handheld sun photometers, for example, are powered with a 9-V battery that can also power the A/D converter, and which provides a common ground for the circuit and the A/D converter.

One additional point about data loggers is worth mentioning. Like all electronic measuring devices, loggers have an "input impedance" that is equivalent to placing a resistance across the output of the circuit being measured. Whether this imposes a significant perturbation on the measurement process depends on the application. Suppose a logger has an input impedance of 10,000 Ω and it is used to record the output from the pyranometer discussed in Chapter 4, with a load resistor of 470 Ω. With the logger attached, its input impedance is in parallel with the 470-Ω load resistance, so that the actual load resistance the photodetector "sees" is now no longer 470 Ω, but

$$R_{load} = (1/470 + 1/10000)^{-1} = 449 \ \Omega$$

Although this does not change the performance of the pyranometer significantly, it does slightly change the calibration and illustrates the point that an instrument and a data collection device must, as a general rule, be considered as a single integrated system. As a practical matter, this might mean that if you use a U12 logger to collect calibration data for a pyranometer, you should use a U12 logger to collect data from that instrument.

Technology in analog-to-digital conversion is advancing rapidly and should make higher resolution data loggers available at lower cost. When you are ready to undertake some of the instrumentation projects in this book which require a data logger, there is no substitute for an online search to see what is currently available.

Appendix 7: Building a Transimpedance Amplifier for Converting Current to Voltage

Solar cells and other photodetectors are current-producing devices. Exposing such devices to light causes electrons to flow in proportion to the incident energy. As previously noted, even light emitting diodes (LEDs) can work this way. In a normal application, applying a voltage to an LED produces light. However, shining light of an appropriate wavelength on an LED will produce a *very* small current compared to even a small solar cell photodetector.

With low-current devices such as LEDs, it is necessary to convert the very small output current to a voltage that can be measured accurately with an inexpensive multimeter or recorded with a data logger. Data loggers, however, require a voltage input, not a current input.[5] In principle, it is possible to convert the current output from a photodetector to an analog voltage output by forcing the device to do "work" (in the physics sense) across a load resistor. This approach was used for the pyranometer described in Chapter 4. As explained in Chapter 4, an important design consideration is that the resistor be small enough to retain the linear relationship between incident solar energy and voltage. The photodetector used for the pyranometer in Chapter 4 produces an output of no more than a few tenths of a volt in full sunlight. With a 12-bit data logger, having a resolution of about 0.6 mV for a 0–2.5 V logger,[6] this is an adequate signal for many purposes. However, the voltage that could be obtained by forcing an LED to do "work" across a load resistor is much too small to be used directly with a data logger.

Connecting a voltage signal that is much smaller than the full-scale range of a data logger won't hurt anything, but at best it may produce an unacceptable loss of resolution in the analog-to-digital conversion of the voltage for computer storage. With a typical LED used as a sunlight detector, you won't detect any usable voltage at all across a load resistor. As a result, it may be desirable or necessary to convert current to a voltage that makes optimum use of the range of the data logger.

---

[5] "Current input" cables available for data loggers actually provide a voltage signal to the logger. They contain a precision resistor that converts a sufficiently large current to a voltage, but this will not work for devices such as LEDs that produce very small currents.

[6] The USB dataloggers from Onset Computer Corporation have a 0–2.5 V range.

To record output from a device that generates a very small current, an operational amplifier can be configured as what is called a transimpedance amplifier—an amplifier that converts current to a voltage that is proportional to the current. This is essentially an "active" version of the "passive" method of forcing a photodetector to develop a potential difference across a resistor. By controlling the gain of such an amplifier, even a very small input current can produce a sizable output voltage. Such a device is easy to build for someone with even basic electronic construction skills, and is usable with all the instruments discussed in this book except the pyranometer, which doesn't need amplification.

Parts list

- Low noise op-amp that will work with a single supply voltage, such as the LTC1050 (Linear Technology Corporation, www.linear.com)
- Feedback resistor, value to be chosen based on the application
- Photodetector
- 0.1 μf capacitor
- Bypass capacitor, value to be chosen based on the application
- SPST on/off switch
- 9-V battery clip
- 9-V battery
- Red and black #22 or #24 gauge wire (solid may be easier to work with than stranded)
- "Perf board" such as RadioShack Part #276–170.

Figure A7.1 shows a circuit schematic for a transimpedance amplifier with a solar cell input. Figure A7.2 shows this circuit assembled on a prototyping board. The rows of give connection points running perpendicular (vertically, in this view) from either side of the center channel are electrically connected, but the points are not connected horizontally. (The layout of this prototyping board match the perf board in the parts list.)

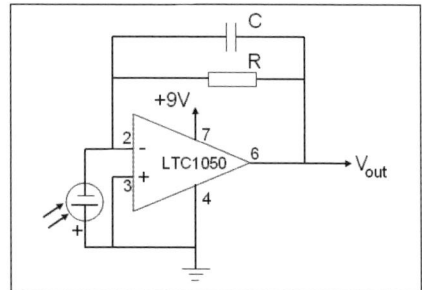

Figure A7.1. Transimpedance amplifier schematic.

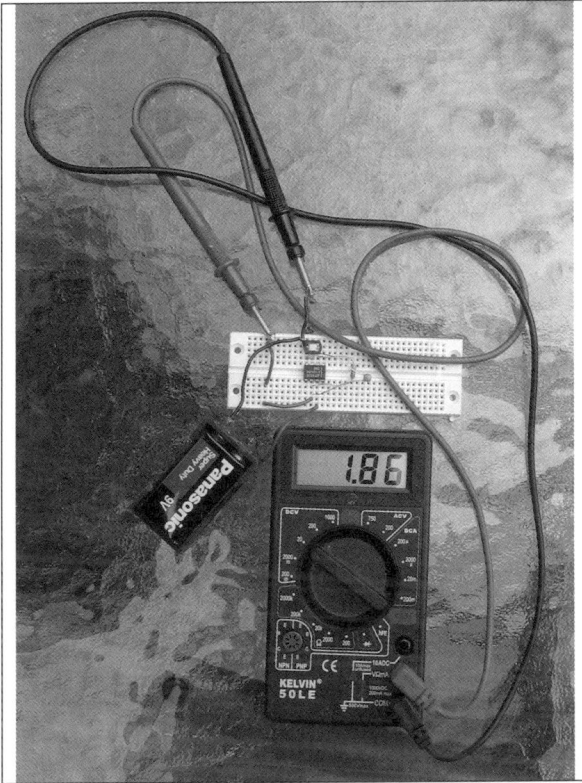

A7.2a. Testing transimpedance amplifier.

A7.2b. Close-up of breadboard amplifier.

Figure A7.2 (see color plates). Building a transimpedance amplifier.

This circuit uses an LTC1050 op-amp, a high-quality low-noise amplifier that, for a DC signal, can operate from a single power supply with pin 4 tied to ground. Although other op amps will work, for these circuits it is not a good idea to use an inexpensive op amp such as the widely available LM741, which requires a double (±) power supply and which will have electronic noise and temperature drift problems.

Op amps must be oriented correctly in a circuit. There is a small semicircular notch in one end of the op-amp case, on the side of the LTC1050 op amp with the white line printed on the case—the right end of the case as seen in Figure A7.2b. The amplifier for this small "button" solar cell uses a 10-KΩ feedback resistor. (The color code is brown-black-orange, reading up from the bottom of the resistor in this view. The gold band at the top indicates that it is a 5% tolerance resistor.) This photodetector was used simply because it was cheap and available, not because it necessarily needs amplification.

The 220-pf capacitor connected in parallel with the feedback resistor serves to stabilize the amplifier, especially when the gain is very high. With some detectors, including some LEDs, the circuit will "self-oscillate" without this capacitor. Probably this is not necessary for this solar cell detector, but it never hurts to include it. The voltage output of this circuit is proportional to the current produced by the solar cell. As shown in Figure A7.2a, the circuit produces almost 2 V under a hazy and overcast summer sky. Such a circuit would no doubt overdrive a 0–2.5 V data logger in full sunlight, which is why the pyranometer described in Chapter 4 does not need a separate amplifier. In the same configuration, an amplifier using an LED as a current-producing device might require a feedback resistor of 10 MΩ or more to get an output of 1–2 V in full sun.

The op amp pins are numbered from 1 through 8, starting in the upper right-hand corner, as positioned in Figure A7.2b and proceeding counterclockwise. The solar cell's "–" lead is on the right-hand side, connected to the op-amp's pin #2. Note the small jumper wire connecting pins #3 and #4. Pin #4 is the circuit ground.

The current drawn by this circuit is very small. It will run continuously for several days even with an inexpensive carbon battery as shown in Figure A7.2a. It is also possible to power this circuit with three or four alkaline or rechargeable AA batteries. For operation just during daylight, you should be able to use a small 6- or 12-V solar panel. Operational amplifiers will work over a wide range of power supply voltages, but bear in mind that the anticipated maximum output voltage should be well under the power supply voltage. It is usual practice to insert a 0.1-μf ceramic disk capacitor across the "+" and "–" leads from the battery. This is to protect the circuit from instability in the power supply, although it is probably not necessary with a battery power supply.

The circuit shown in Figure A7.2 can be reproduced on perf boards that are configured identically to the prototyping board.[7] A very compact version of this amplifier can be built on a much smaller piece of perf board, as shown in Figure A7.3.[8] It lacks only the power supply connections, to be made on the right-hand edge of the perf board. It will fit inside a piece of 3/4" Schedule 80 PVC plumbing pipe.

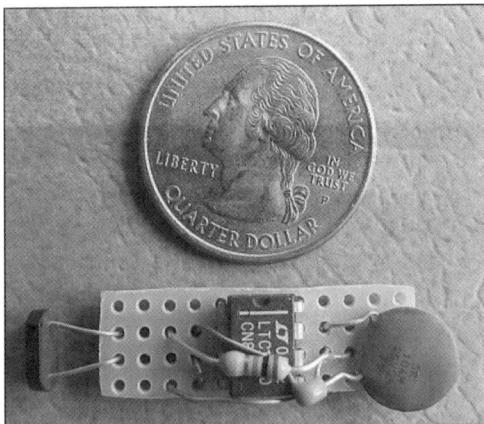

Figure A7.3. A small version of the transimpedance amplifier with a solar cell detector.

With LED detectors, the feedback resistance will be much higher than the 10 kΩ used in this solar-cell circuit. For some LEDs, 10 MΩ or more will be required. High gain usually is not a problem, although it is possible that such a circuit will generate noticeable electronic noise or may become unstable. These kinds of problems are difficult to track down with just a multimeter. For example, if you are expecting a DC output voltage from your transimpedance amplifier, a "DC volts" setting on a multimeter will simply average out any AC component that might be present. Even a circuit that is oscillating wildly may still produce an averaged DC output that looks reasonable.

The only reliable way to check the stability of these kinds of circuits is to use a laboratory oscilloscope. This is easy to do, but is complicated by the fact that the detectors in these circuits need to be exposed to sunlight, and it is not practical to replicate this condition under indoor lighting. There are a few sources of portable handheld battery-powered portable oscilloscopes, which are an ideal solution for working with these instruments.

---

[7] For example, RadioShack Part #276-170.

[8] "Surface mount" components are even smaller, but they are much more difficult to work with.

Appendix 8: Calculating the Ångstrom Exponent and Turbidity Coefficient

Optical thickness is defined by measuring the amount of direct sunlight reaching a detector that responds (theoretically) to a single wavelength of light. Aerosol optical thickness (AOT) is that portion of optical thickness due to aerosols, α_λ. AOT, wavelength, and atmospheric turbidity, β (haziness) are related through Ångstrom's[9] turbidity formula:

$$\alpha_\lambda = \beta \bullet \lambda^{-\tau} \tag{A8.1}$$

where β is the Ångstrom turbidity coefficient, λ is wavelength in microns, and τ is the Ångstrom exponent. τ and β are independent of wavelength, and can be used to describe the size distribution of aerosol particles and the general haziness of the atmosphere. For two different wavelengths,

$$\alpha_{\lambda,1} = \beta \bullet \lambda_1^{-\tau} \tag{A8.2a}$$
$$\alpha_{\lambda,2} = \beta \bullet \lambda_2^{-\tau} \tag{A8.2b}$$

from which

$$\alpha_{\lambda,1}/(\lambda_1^{-\tau}) = \alpha_{\lambda,2}/(\lambda_2^{-\tau}) \tag{A8.3}$$

Solving for τ:

$$\tau = \ln(\alpha_{\lambda,1}/\alpha_{\lambda,2})/\ln(\lambda_2/\lambda_1) \tag{A8.4}$$

A typical range for τ is 0.5–2.5, with an average for typical atmospheres of around 1.3 ± 0.5. Larger values of τ, when the AOT value for the larger wavelength is much smaller than the AOT value for the smaller wavelength, imply a relatively high ratio of small particles to large ($r > 0.5\ \mu$) particles. As AOT for the larger wavelength approaches AOT for the smaller wavelength, larger particles dominate the distribution and τ gets smaller. It is not physically reasonable for the AOT

---

[9] This value is named after Anders Jonas Ångstrom, a 19th century Swedish physicist considered one of the founders of the field of spectroscopy.

value of the larger wavelength to equal or exceed the AOT value of the smaller wavelength.

Now calculate β from either wavelength:

$$\beta = \alpha_{\lambda,1} \cdot \lambda_1{}^{\tau} = \alpha_{\lambda,2} \cdot \lambda_2{}^{\tau} \tag{A8.5}$$

where λ must be expressed in microns (500 nm = 0.500 μ). Values of β less than 0.1 are associated with a relatively clear atmosphere, and values greater than 0.2 are associated with a relatively hazy atmosphere.

Given AOT at two different wavelengths, AOT at a third wavelength can be inferred for the same atmospheric conditions. Rewrite (A8.5) and solve for $\alpha_{\lambda,3}$ using either the first or second wavelength:

$$ln\,(\lambda_3/\lambda_1)\tau = ln\,(\alpha_{\lambda,1}/\alpha_{\lambda,3}) = ln\,(\alpha_{\lambda,1}) - ln\,(\alpha_{\lambda,3}) \tag{A8.6}$$
$$ln\,(\alpha_{\lambda,3}) = ln\,(\alpha_{\lambda,1}) - ln\,(\lambda_3/\lambda_1)\tau$$
$$\alpha_{\lambda,3} = \exp[ln\,(\alpha_{\lambda,1}) - ln\,(\lambda_3/\lambda_1)\tau]$$

This calculation is useful when AOT values determined with one instrument must be compared to values from another instrument that uses different wavelengths.

Here is a worked-out example for wavelengths used in the two-channel sun photometer described in Chapter 5:

$\lambda_1 = 505$ nm, $\lambda_2 = 625$ nm
$\alpha_{\lambda,1} = 0.185$, $\alpha_{\lambda,2} = 0.155$
$\tau = ln\,(0.185/0.155)/ln\,(625/505) = 0.8299$

Using the first wavelength, $\beta = 0.185 \cdot 0.505^{0.8299} = 0.1049$

Find AOT for a wavelength of 550 nm:
$\alpha_{550} = \exp[ln\,(0.185) - ln\,(550/505) \cdot 0.8299] = 0.1723$

More information about these calculations can be found at WMO [1990] or Iqbal [1983]. (Iqbal's very useful book should be available in large technical libraries, but it has long been out of print.)

References

Abbott, C. G. and F. E. Fowle, Jr., *Annals of the Astrophysical Observatory of the Smithsonian Institution*, Vol. II, Part 1, 11-124. US GPO, Washington, DC, 1908.

Allison, Mead A., Arthur T. DeGaetano, and Jay M. Pasachoff. *Earth Science* , Holt, Rinehart and Winston, 2006.

American Chemical Society, *Chemistry in Context: Applying Chemistry to Society* , 2000.

Apogee Instruments, http://www.apogeeinstruments.com/pdf_files/PYRManual.pdf, 2008.

Bird, R. E., and R. L. Hulstrom, Simplified Clear Sky Model for Direct and Diffuse Insolation on Horizontal Surfaces, Solar Energy Research Institute Technical Report No. SERI/TR-642-761, Golden, CO, 1981.

Bird, R. E., and C. J. Riordan, Simple Solar Spectral Model for Direct and Diffuse Irradiance on Horizontal and Tilted Planes at the Earth's Surface for Cloudless Atmospheres. *Journal of Climate and Applied Meteorology*. Vol. 25(1), January 1986; pp. 87-97, 1986.

Boersma, K. F., and J. P. de Vroom. Validation of MODIS Aerosol Observations over the Netherlands with GLOBE Student Participation. *Journal of Geophysical Research*, **111**, *D20*, D20311, 2006.

Brooks, David. R. and F. M. Mims III, Development of an inexpensive handheld LED-based Sun photometer for the GLOBE program, *Journal of Geophysical Research, Res.* **106**, *D5*, 4733-40, 2001.

Brooks, David R., Forrest M. Mims III, Arlene S. Levine, and Dwayne Hinton: The GLOBE/GIFTS Water Vapor Monitoring Project: An Educator's Guide with Activities in Earth Sciences, National Aeronautics and Space Administration, EG-2003-12-06-LARC, 2003a.

Brooks, David R., F. Niepold, G. D'Emilio, J. Glist, G. Hatterscheid, S. Martin, K. Dede, and I. Neumann, Scientist-Teacher-Student Partnerships for Aerosol Optical Thickness Measurements in Support of Ground Validation Programs for Remote Sensing Spacecraft, IAC-03-P.4.07, International Astronautical Federation, 54th International Astronautical Congress, Bremen, Germany, Sept. 28 – Oct. 3, 2003b.

Brooks, David R., F. M. Mims III, and R. Roettger, Inexpensive Near-IR Sun Photometer for Measuring Total Column Water Vapor, *Journal of Atmospheric and Oceanic Technology,* **24**, 1268-76, 2007.

Bucholtz, Anthony, Rayleigh-Scattering Calculations for the Terrestrial Atmosphere, *Applied Optics*, **34**, *15*, 2765-73, 1995.

Cannon, Annie Jump, and Edward Charles Pickering, *Annals of the Astronomical Observatory of Harvard College,* **56**, *4*, Cambridge, Mass., 1912.

Diak, G. R., W. L. Bland, J. R. Mecikalski, *et al.*, A Note on First Estimates of Surface Insolation from GOES-8 Visible Satellite Data, *Agricultural and Forest Meteorology,* **82**, 219-226, 1996.

Duchon, C. E., M. S. O'Malley, Estimating Cloud Type from Pyranometer Observations, *Journal of Applied Meteorology*, **38**, 132-141, 1999.

Einstein, Albert, Über einen die Erzeugung und Verwandlung des Lichtes betreffenden heuristischen Gesichtspunkt [On a Heuristic Point of View about the Creation and Conversion of Light], *Annalen der Physik*, Leipzig, **17**, 132, 1905.

Gates, D. M., and W. J. Harrop, Infrared transmission of the atmosphere to solar radiation. *Applied. Optics*, **2**, 887-898, 1963.

Gueymard, C.A., SMARTS, A Simple Model of the Atmospheric Radiative Transfer of Sunshine: Algorithms and Performance Assessment, Technical Report No. FSEC-PF-270-95. Cocoa, FL: Florida Solar Energy Center, Cocoa, FL, 1995.

Gutman, S., and K. Holub, Ground-Based GPS Meteorology at the NOAA Forecast Systems Laboratory, *Preprints, Fourth Symposium on*

Integrated Observing Systems, Long Beach, American Meteorological Society, 1-5 , 2000.

Holben B. N., T. F. Eck, I. Slutsker, D. Tanré, J. P. Buis, A. Setzer, E. Vermote, J. A. Reagan, Y. Kaufman, T. Nakajima, F. Lavenu, I. Jankowiak, and A. Smirnov, AERONET—A Federated Instrument Network and Data Archive for Aerosol Characterization, *Remote Sensing of the Environment,* **66**, 1-16,1998, http://aeronet.gsfc.nasa.gov/

Iqbal, Muhammad. *An Introduction to Solar Radiation*, Academic Press, Toronto, 1983

King, David L., and Daryl R. Myers, Silicon-Photodiode Pyranometers: Operational Characteristics, Historical Experiences, and New Calibration Procedures, 26th IEEE Photovoltaic Specialists Conference, September 29 – October 3, 1997, Anaheim, California, pp. 1285-1288.

Leckner, B., The Spectral Distribution of Solar Radiation at the Earth's Surface—Elements of a Model, *Solar Energy*, Vol. 20, 143-150, 1978.

Mims, Forrest M., III, Sun Photometer with Light-Emitting Diodes as Spectrally Selective Detectors. *Applied Optics*, **31**, *33*, 6965-67, 1992.

Mims, Forrest M. III, An Inexpensive and Stable LED Sun Photometer for Measuring the Water Vapor Column over South Texas from 1990 to 2001, *Geophysical Research. Letters*, **29**, *13*, 20-1—20-4, 2002.

Mims, Forrest M. III, A 5-Year Study of a New Kind of Photosynthetically Active Radiation Sensor. *Photochemistry and Photobiology*, **77**, *1*, 30-33, 2003a.

Mims, Forrest M. III, *Getting Started in Electronics*, Master Publishing, Lincolnwood, IL, 2003b. ISBN 0945053282.

Meeus, Jean, *Astronomical Algorithms*, Willmann-Bell, Richmond, Virginia, 1991. ISBN 0-943396-35-2.

NASA, 2008. http://modis-atmos.gsfc.nasa.gov

NASA, May 2001 (last update).
http://www.nas.nasa.gov/About/Education/Ozone/history.html

National Renewable Energy Laboratory, 2008.
http://www.nrel.gov/rredc/smarts/references.html

National Research Council, *National Science Education Standards*, 1996.
ISBN 0-309-05326-9. http://www.nap.edu/readingroom/books/nses/

NOAA, 2007. http://www.srh.noaa.gov/elp/wxcalc/formulas/wetBulbTd
FromRh.html

Planck, Max, On the Law of Distribution of Energy in the Normal
Spectrum, *Annalen der Physik*, **4**, 553, 1901.

Reitan, C. H., Surface Dew Point and Water Vapor Aloft, *Journal of
Applied Meteorology* **2**, 776-79, 1963.

Smith, W. L., Note on the Relationship Between Total Precipitable Water
and Surface Dew Point. *Journal of Applied Meteorology*, **5**, 726-727,
1966.

Tregoning, P., R. Boers, D. O'Brien, and M. Hendy, Accuracy of
Absolute Precipitable Water Vapor Estimates from GPS Observations.
Journal of Geophysical Research, **103**, 28701–28710, 1998.

Voltz, Frederick E., Economical Multispectral Sun Photometer for
Measurements of Aerosol Extinction from 0.44 μm to 1.6 μm and
Precipitable Water, *Applied Optics*, **13**, 1732-33, 1974.

WMO, **B**ackground **A**ir **P**ollution **MON**itoring (BAPMON) Network
Information Manual, TD-9789, September, 1990.

Wood, R. W., Note on the Theory of the Greenhouse, *Edinborough and
Dublin Philosophical Magazine*, **17**, 319-320, 1909.

Young, Andrew T., Air Mass and Refraction, *Applied Optics*, **33**, *6*,
1108-10, 1994.

Color Plates

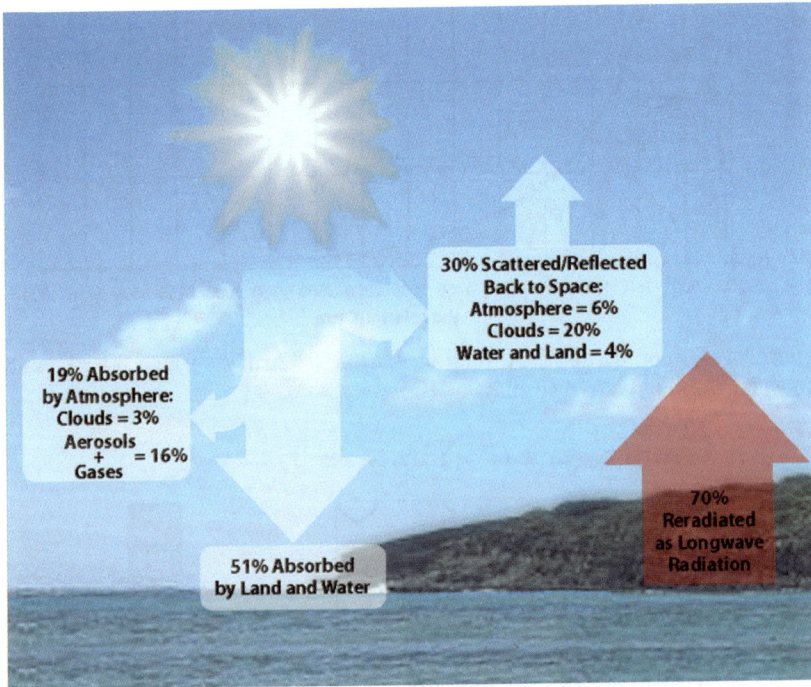

Figure 1.1. Schematic representation of Earth's radiative balance [Graphic by Vivek Dwivedi, NASA Goddard Space Flight Center].

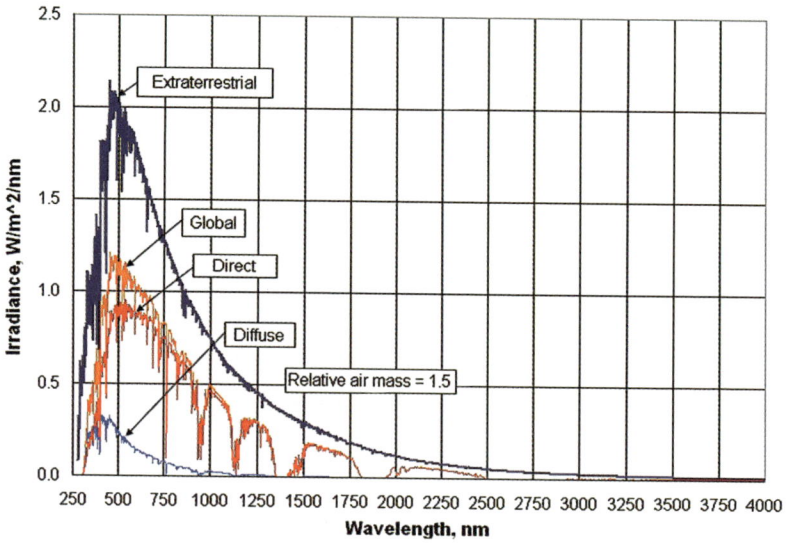

Figure 3.1. Direct, diffuse, and total insolation at Earth's surface for a standard atmosphere and a relative air mass of 1.5.

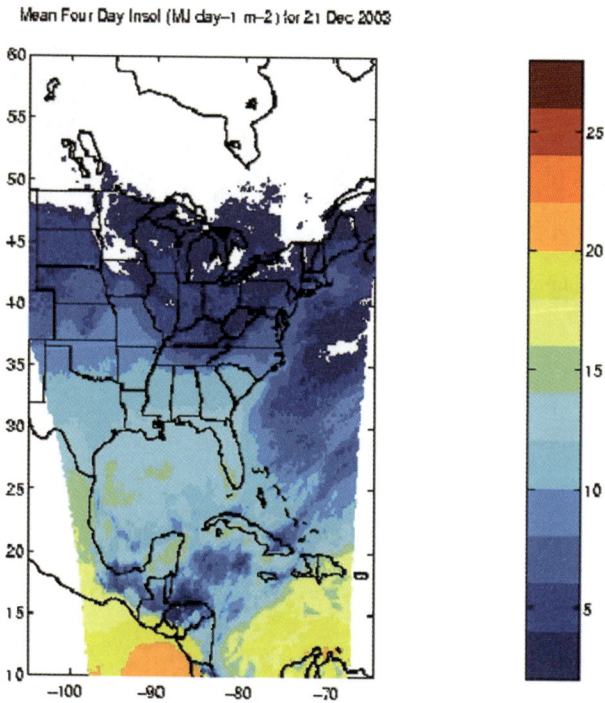

Mean Four Day Insol (MJ day–1 m–2) for 21 Dec 2003

Figure 3.6. 4-day mean insolation over North America, late December 2003, based on GOES visible images [Diak *et al.*, 1996].

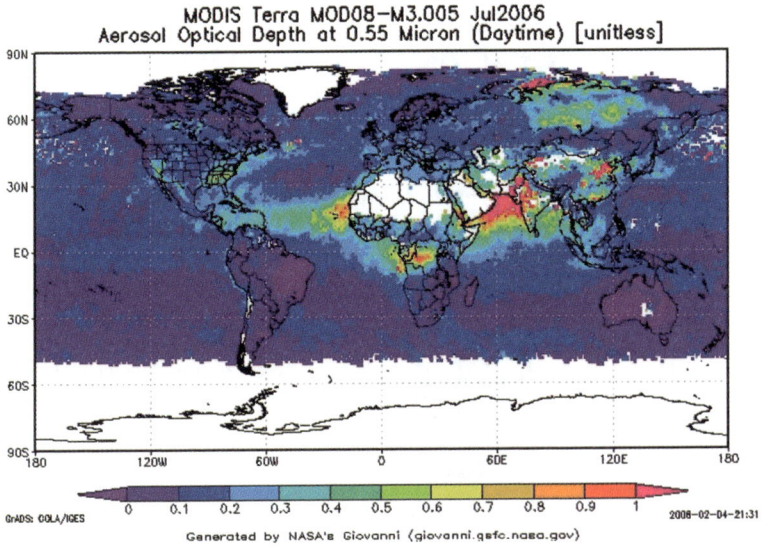

Figure 3.11. 550-nm aerosol optical depth from MODIS/Terra, monthly mean values for July, 2006 [See http://modis-atmos.gsfc.nasa.gov/].

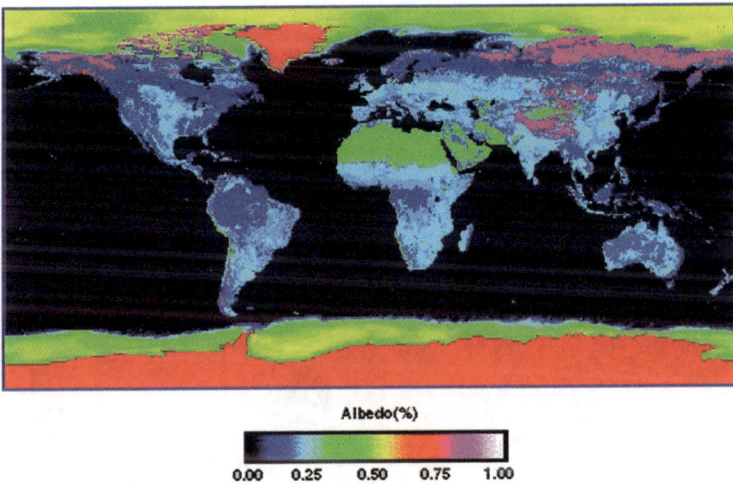

Figure 3.16. False-color representation of monthly average broadband albedo, October, 1986 [www-surf.larc.nasa.gov/surf/pages/bbalb.html].

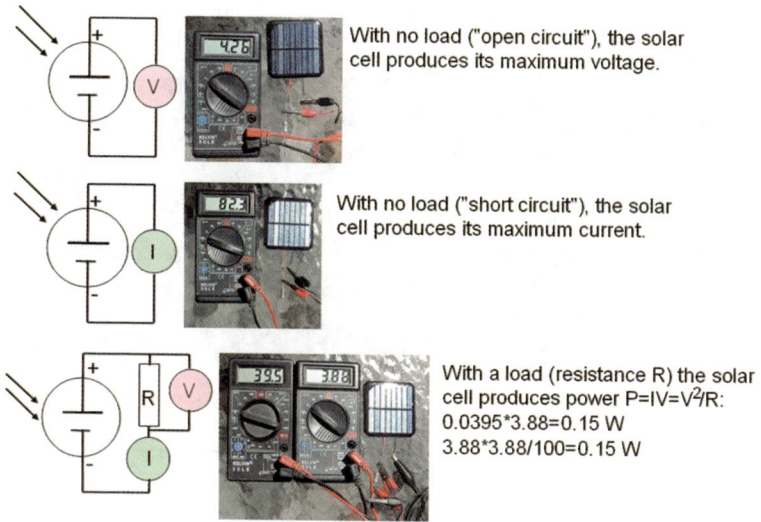

With no load ("open circuit"), the solar
cell produces its maximum voltage.

With no load ("short circuit"), the solar
cell produces its maximum current.

With a load (resistance R) the solar
cell produces power P=IV=V^2/R:
0.0395*3.88=0.15 W
3.88*3.88/100=0.15 W

Figure 4.3. Measurements on a solar cell: open-circuit voltage, short-circuit current, and work across a resistor.

Figure 4.5. Current output from PDB-C139 silicon photodiode with diffuser.

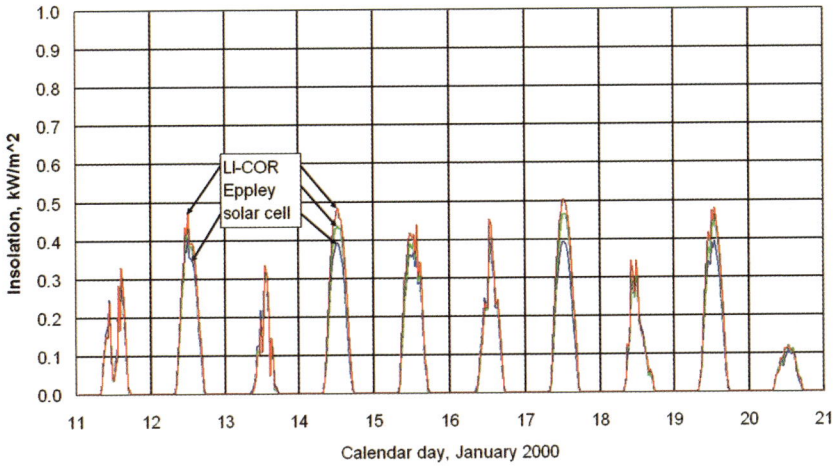

4.9a. Solar insolation during January 2000.

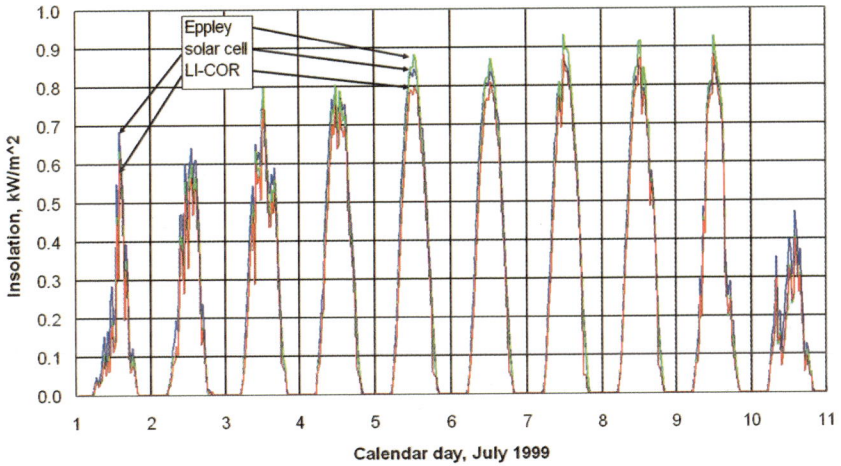

4.9b. Solar insolation during July 1999.

Figure 4.9. Solar insolation comparisons with three different pyranometers, Philadelphia, Pennsylvania, USA (lat = 39.96 N, lon = 75.19 W).

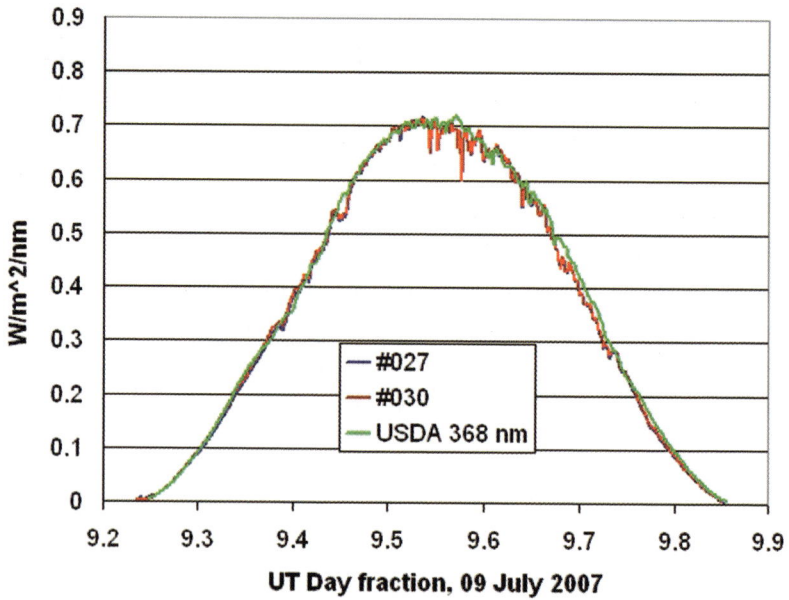

4.20a. No cosine correction.

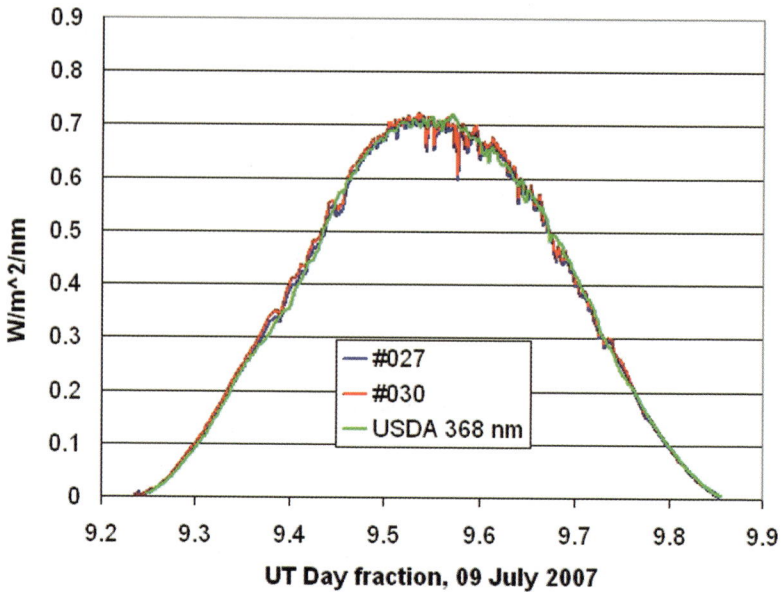

4.20b. With cosine correction.

Figure 4.20. Two UV-A radiometers calibrated against a USDA UV site in Beltsville, Maryland, without and with cosine correction.

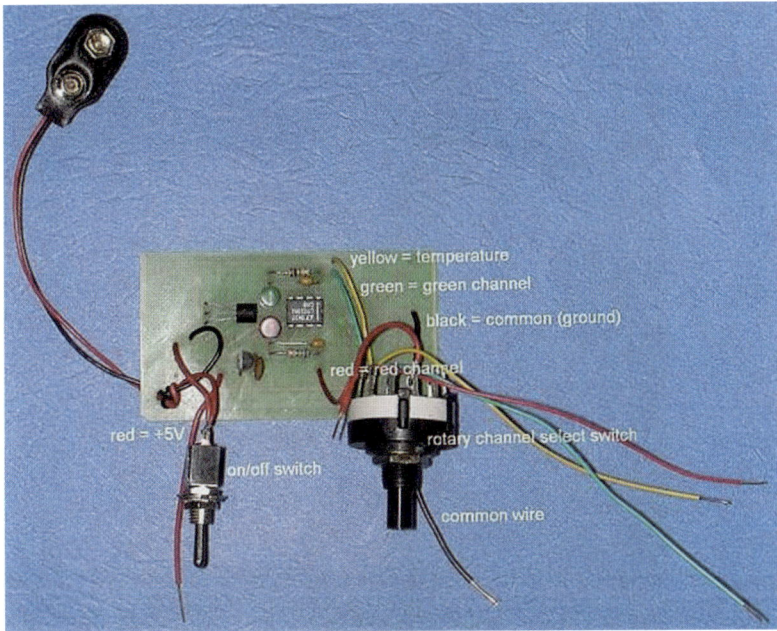

Figure 5.3. Printed circuit board assembly for two-channel sun photometer.

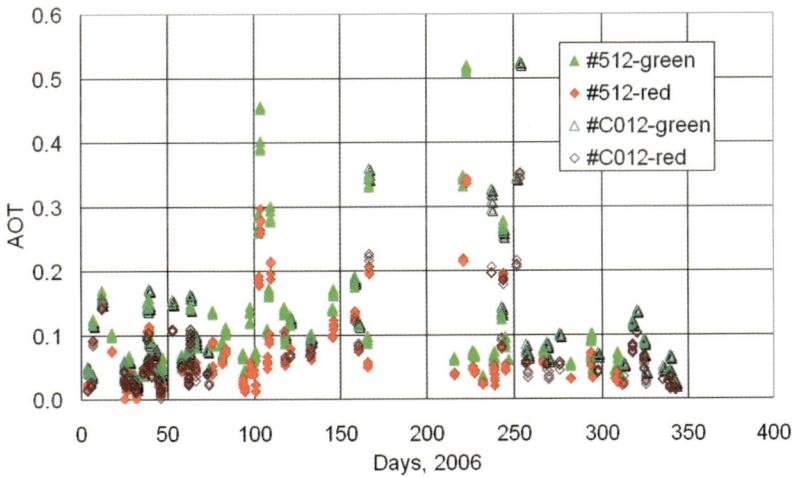

Figure 5.6. Aerosol optical thickness data from two sun photometers at a rural school in Arkansas. Data provided by Wade Geery. Latitude 36.1972° N, longitude 92.2688° W.

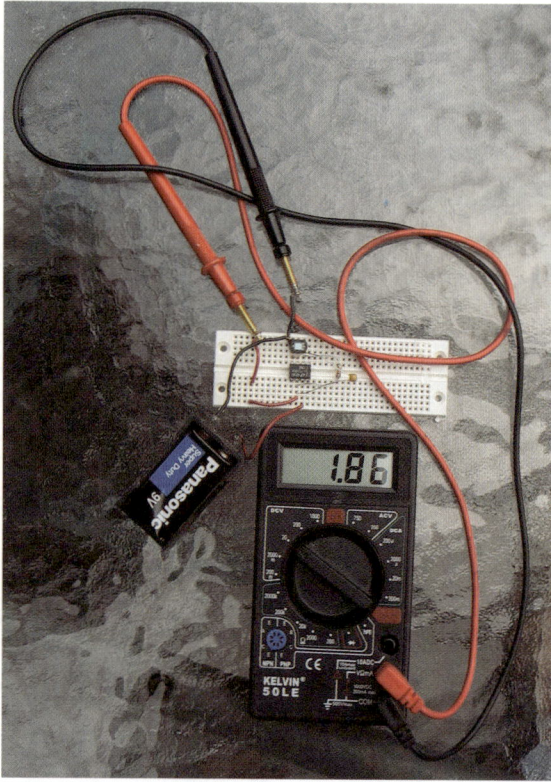

A7.2a. Testing transimpedance amplifier.

A7.2b. Close-up of breadboard amplifier.

Figure A7.2. Building a transimpedance amplifier.

Index